Sustainable Futures for Climate Adaptation

Considering sustainability as a flawed and restrictive term in practice, *Sustainable Futures for Climate Adaptation* argues that we must radically adapt humanity and reform society, cities, buildings, and our approach to migration in order to coexist in harmony with our natural environments.

The book conceives an Earth–human coexistence where the world's regions are shared globally between all people, in contrast to a reality where we have lost touch with the natural world. It is this decoupling of humanity and nature that has brought us to the brink of climate disaster. In response, Benedict Anderson explores the concept of 'wearing our ecology', where human mobility is synchronized with the environment, merging people with landscapes, topographies, and geographies. Anderson argues that we need to create new migration routes for people moving between the Global South and North and establish flexible and adaptive living environments. Only by rethinking separations between urban and rural, resource extraction and consumption, racial prejudice and accessibility are we able to forge a closer partnership with nature to adapt to climate change and mitigate the worst of its effects.

Touching on themes of adaptive urban design, racial and gender segregation and inequality, and climate apocalypticism, this book will be valuable reading for researchers, scholars, and upper-level students in the fields of urban studies, migration studies, human geography, ecology, politics, and design.

Benedict Anderson is an independent scholar and practices in design, architecture, and public art. He has held academic and professorial positions in many different universities, lectured extensively as an invited speaker, and exhibited in major exhibitions around the world. His previous books for Routledge are *Buried City, Unearthing Teufelsberg: Berlin and Its Geography of Forgetting* (2017), *The City in Geography: Renaturing the Built Environment* (2019), and *The City in Transgression: Human Mobility and Resistance in the 21st Century* (2020).

Sustainable Futures for Climate Adaptation

Wearing Our Ecology

Benedict Anderson

Routledge
Taylor & Francis Group
LONDON AND NEW YORK

earthscan
from Routledge

Designed cover image: Gas flaring, Nahran Omar oil field, Iraq. Image still from documentary film *Under Poisoned Skies*, director Jess Kelly, producer Owen Pinnell, BBC Arabic Investigations, 2022.

First published 2024
by Routledge
4 Park Square, Milton Park, Abingdon, Oxon OX14 4RN

and by Routledge
605 Third Avenue, New York, NY 10158

Routledge is an imprint of the Taylor & Francis Group, an informa business

British Library Cataloguing-in-Publication Data
A catalogue record for this book is available from the British Library

Library of Congress Cataloging-in-Publication Data
Names: Anderson, Benedict, author.
Title: Sustainable futures for climate adaptation :
wearing our ecology / Benedict Anderson.
Description: First edition. | New York : Routledge, 2024. |
Includes bibliographical references and index.
Subjects: LCSH: Global environmental change. |
Global temperature changes. | Sustainable development. |
Conservation of natural resources. |
Population geography. | Economic geography.
Classification: LCC GE149 .A53 2024 (print) |
LCC GE149 (ebook) | DDC 304.2/8–dc23/eng/20231012
LC record available at https://lccn.loc.gov/2023037326
LC ebook record available at https://lccn.loc.gov/2023037327

ISBN: 978-1-032-43862-7 (hbk)
ISBN: 978-1-032-46604-0 (pbk)
ISBN: 978-1-003-38251-5 (ebk)

DOI: 10.4324/9781003382515

Typeset in Sabon
by Newgen Publishing UK

Contents

Figures

Preface

Sweltering in my home in Merida, Mexico, where I spend half the year, the heat is becoming more intense, the air more humid every year. On the other side of the world in Berlin, where I spend the other half of the year, it is also getting warmer. During the winter it doesn't snow much anymore and it's not as cold. When I first came to Berlin in the early 1990s, the winter would hover around 0 to -10°C or lower, the type of cold that freezes your ear lobes. It still gets cold, but it now hovers between -5 and +7 or higher. Spring tends to now come what seems to be overnight and the summer is earlier and longer. These seasonal changes appear to cause confusion not only to Berliners but to birds, trees, and flowers. Last summer I bought a fan—I had never needed one before. I work in the attic of my apartment block, a room with a view of roofs, chimneys, and sky that I pretentiously call my Heidegger's Hut for its natural pine-lined interior. It becomes unbearably hot to work there now and, as far I can recall, it never used to be that way.

I come from the driest continent on Earth: Australia. In the last two decades, extreme weather events have become more prevalent than ever before, with floods, fires, drought, soil erosion, crop failure, water scarcity, and coral bleaching causing catastrophic environmental destruction and extinction of native species. This is a country where the mining of minerals is widespread. In 2022, Australia dug up 550 million tons of coal to power its electricity stations and export to countries such as China to power theirs. Growing up in the farming region and semi-desert on the west coast of South Australia, my memories are of dusty tracks, glaring sun, and hot dry winds that blow away the top soil from bare treeless paddocks. Not until my early teens did I learn about the benefits of block-out sun cream in a country where skin cancer is rife. I remember when plastic bags were introduced at supermarkets and their first appearances after being washed-up along the beach.

In the timeframe of the earth's 4.55 billion years of existence, human existence is a minuscule dot in comparison. In the last 200 years, humans have undertaken mass excavation of resources, industrial and technological revolution—forging a perception and practice of dominance over the earth. Polluting emissions from factories have resulted in atmospheric turbulence causing devastating weather events and extreme hardship for billions of people across the world. Since I was born, countless wars and famines have raged across the planet. The legacies of the European colonial era still persist as does racial, gender, political, and cultural discrimination of all kinds. Forests the size of football fields are cleared every hour; glaciers and sea ice are melting at an alarming rate; and, if this were not frightening enough, 150–200 species of animals and plants become extinct each day. Climate change is now on most people's minds. Global warming is indisputable and climate deniers continue to push their rhetoric that it is a hoax. Dedicated militant activists and angry teenagers who will inherit the earth have taken to the streets calling for climate action. Fossil fuel companies are the enemy and green energy, such as electric cars, solar panels, wind farms, better water and land management, and education about the causes of global warming, is the friend.

Where climate disruption oscillates in the earth's atmosphere causing catastrophic weather events on the ground, the impact of such events is not evenly spread. One quarter of the world's population, two billion people, face water and food insecurity on a daily basis due to climate change. Many of these people reside in underdeveloped and former colonial countries collectively referred to as the 'Global South'. In contrast, the wealthy industrial countries of the 'Global North', who are responsible for most of the world's greenhouse gas emissions that cause global warming, are significantly less affected due to their extensive infrastructure and financial resources. The wealth gap between rich and poor countries increases year-on-year as does global military spending. If the finances allocated to military spending in 1 year were redirected to climate change adaptation programmes, the funds would support these programmes for the next 10 years. The attitudes and imbalance of priorities to address climate change speak to the absurdity, reality, fear, anger, and denial that seesaw in the collective psyche of the human condition and between the positive and negative actions of governments, people, and powerful corporations.

I do not believe that this book will have any real effect in providing answers or solutions to climate change and global sustainability—what can a book offer to the most pressing issue facing humankind? But I hope it might do something in terms of how humanity can rethink its relations to the natural world. Through a reappraisal of human history to the present and into the future, this book offers some ideas as to how humans might restructure habitation, combat inequality, and better comprehend animal and plant life to inform how to live and adapt to climate change. Throughout writing this book, I have often felt that my time would be better spent being a full-time activist in pressuring governments, institutions, and corporations for change and to dismantle the regimes of capital and exploitation that have propped-up the world order and inequity. Whatever humans do alters or damages the environment. I have come to understand that a global sustainable future and global equality must go hand in hand. As humans continue to exploit the earth's resources at an exponentially increasing rate, they further jeopardize their own existence and that of all living things. I started off living in a world where anything goes; environmental abuse and seemingly infinite consumption were rife. I now live in a world threatened by environmental collapse. Out on a clear blue sky, it is easy to disbelieve the effects of climate change. To the reality of hundreds of millions of people across the world forced to leave their homelands due to climate change and suffering famine, belief and collapse has already come.

Acknowledgements

This is my fourth book exploring themes on the city, geography, human mobility to include issues surrounding climate change, sustainability, and more broadly the future of human and Earth coexistence. As I note in the introduction, this is a book I initially did not want to write; too vast was the subject and too prone to failure. But I have come to understand that writing is a combination of failure and success; it is elusive as much as defining meaning behind thoughts and ideas.

I have a few people and organizations to thank who have generously supplied various images contained in the book. They include in no particular order: Jessica Kelly for her image of gas flaring at the Nahran Omar oil field in Iraq from her film *Under Poisoned Skies*, NASA satellite images such as the image showing the atmospheric turbulence of Hurricane Ian, the not-for-profit organization Decolonial Atlas, specifically the work of Jordan Engel who, with fellow cartographers, have managed to create a vast library of cartogram maps illustrating, among many themes, global inequality, plastic pollution, and the countries, companies, and individuals responsible for global CO_2 emissions and environmental devastation. I highly recommend viewing their work. I would also like to thank Sophie Chao for lending me a drone image of deforestation in Papua New Guinea, OTSkydrone for their drone images of NEOM project showing excavation earthworks in Saudi Arabia and the desert highway, and the public platform Pexels. For images by Vlad Chețan and others that I have used, I am very grateful. I would also like to thank Climeworks for the image showing their carbon-capturing plant in Iceland, and Nancy Diniz in providing the image displaying the technology for reading PM2 fine-particle air pollution in our collaborative project 'Wearing Your Air' at the Lisbon Architecture Triennale. Last but not least, I'd like to thank Simon Mussell who, as with two of my previous books, edited this volume. I am unsure how much pain and suffering he goes through while editing my sometimes-wayward lines, and though he assures me there is no pain, I feel he is being polite. At Routledge, I would like to thank commissioning editor Grace Harrison, whom I have previously worked with on first book *Buried City*, for once again being supportive in seeing the value of this fourth book and editor Matthew Shobbrook in seeing this book through to completion. I would also like to thank project manager Aloysias Saint Thomas, copyeditor Smita Agarwal, and indexer Damian Love.

As with my other books, I wrote this book in Mexico and Berlin in two climate extremes of hot and cold. Coming from the driest state on the driest continent on Earth, Australia, having witnessed sandstorms, treeless landscapes, and cracked earth from long periods of drought, I do not claim to know more than anyone else concerning the multifarious complexities of climate change. What I do know is that I am living through major environmental devastation, mass species extinction, extreme weather events due to catastrophic climatic conditions as a result of human impact. I have experienced extreme weather turbulence over the decades of my life, and I know that, like others living in a privileged Western society surrounded by unlimited consumption, I cannot escape responsibility for the catastrophic weather events engulfing the earth and affecting the lives of billions of people and all living things.

Introduction
Earth Overshoot

Nature is humanity's best friend. Without nature we have nothing. Without nature we are nothing... And yet humanity seems hell-bent on destruction. We are waging war on nature. Deforestation and deacidification are creating wastelands of once thriving eco-systems... Multinational corporations are filling their bank accounts while emptying our world of its natural grifts... With our bottomless appetite for unchecked and unequal economic growth, humanity has become a weapon of mass extinction. We are treating nature like a toilet and ultimately, we are committing suicide by proxy... And the cost we measure of the indifferent and unjust and the culpable losses to the poorest countries, indigenous populations, women and young people. Those least responsible for this destruction are always the first to feel the impacts but they are never the last. Dear friends this conference is on a chase to stop this orgy of destruction, to move from discord to harmony and to apply the ambition and action to challenge demands.

(Edited excerpt from United Nations General Secretary Antonio Guterres' opening address at the COP15 United Nations Biodiversity Conference in Toronto, 6 December 2022)[1]

The effects of climate change are being felt worldwide. Weather turbulence and environmental destruction are becoming ever more common. Every other day while writing this book, reports of catastrophic weather events from around the world were beamed onto TV channels, printed in newspapers, discussed on the radio, and captioned in internet feeds. The aforementioned edited transcript from UN General Secretary Antonio Guterres' address at the 2022 COP15 Biodiversity Conference is one of many he has given to call attention to the most urgent issue facing humankind. 'Nature is humanity's best friend. Without nature we have nothing', he says. The focus of that conference was to reach a consensus between the 196 participating countries to agree to preserve global biodiversity by establishing 30% (recognized as the bare minimum) of the planet's oceans and lands as protected reserves by 2030. A month before, Guterres addressed the 2022 COP27 conference in

DOI: 10.4324/9781003382515-1

Sharm El Sheikh, Egypt, where he warned his audience: 'The clock is ticking; we are in the fight of our lives...we are on a highway to climate hell with our foot still on the accelerator'. Guterres has been an outspoken critic of governments' and fossil fuel companies' continued flagrant obligations to address climate change by placing growth and profit before the health of the world and the lives of future generations.

Alongside daily reports of catastrophic weather events, every other month climate change summits were being held across the world to raise the emergency alarm and put pressure on governments and fossil fuel companies to rein in global carbon emissions. Somewhere around the world, climate change activists were lying down on highways disrupting traffic, deflating the tyres of SUVs, sabotaging fossil fuel mining machinery, chaining themselves, and gluing their hands to the entrances of banks accused of financing fossil fuel companies and covering paintings depicting idyllic rural scenes of the past with dystopian scenes of the future. All faced criticism and some criminal proceedings for their disruptive actions to bring awareness and save the planet from further environmental collapse. All these events and reports, conferences, and actions affected the writing of this book, at times making what was current on the day obsolete the next. It was impossible to keep up, cite, and analyse the fast-changing movements across the world to all the facets of climate change. Halfway through the book, I decided to refocus having reached the conclusion that writing a book on sustainability and climate change was essentially a flawed exercise. For one thing, it was too big an undertaking to fully comprehend and gather all relevant information and to avoid falling into journalistic reportage on economic policies, sustainable practices, national and international projects, which the media does far better. So, I decided to focus on the human costs of climate change: the effects of destructive weather events on people, homes, livelihoods, and survival. I also wanted to understand what I saw as the estrangement between humanity, especially urban dwellers, and the natural world by following the historical course of human settlement through to the industrial and technological ages. There was also a coming to terms with humanity's obsession with shaping the earth according to its desire and placing all living things under its control to service its needs. By exploring these themes, I might provide some answers that have led up to the climate crisis and point to some avenues for departure from this history and ways to relink human habitation and nature.

To write about climate change, human impact on the earth, atmospheric turbulence, sustainability, environmental adaptation, government policies for greenhouse reduction targets, fossil fuel corporations and profit, land and ocean degradation, the impact of war, drought, famine, migration, and global economic and gender inequality, all of which have to be addressed to some degree, felt overwhelming. Seeking information from scientific reports, statistics, and global agreements between nations is a part of understanding the ramifications climate change is exerting on the world. There is also the possibility of getting bogged down in citing an inexhaustible amount of information at the cost of moving freely across ideas concerning climate change and humanity. One constant message I became aware of was the inconsistency of humanity's ability to tackle climate change and achieve global sustainability. Restoring the earth's biosphere and ecologies, seas, and continents, and equalizing opportunities between peoples and between nations appeared too hard in terms of global cooperation. As mentioned in the preface, my time would perhaps have been more effectual by joining full-time groups such as Extinction Rebellion, Just Stop Oil, Fridays for Future, and Last Generation rather than writing this book. It has not been easy; it remains unfinished, flawed, and incomplete—something I knew it was always going to be from the start.

This book ranges widely. There is no real summary as summaries go while so much about the future concerning the impact of climate change, humankind, and the earth is uncertain. To tackle a vast subject that spans multiple disciplinary areas concerning sustainability in an age of planetary crisis is a fool's errand. It is foolish to think that a book can be persuasive to offer pathways out of the present climate crisis, to give solutions about how humans can reduce their impact on the earth, survive global weather turbulence, remove global inequality, and take down the dominance of capitalist systems that have controlled and carved up the world. Then there is the question of how to address other important issues such as the rise of climate change refugees who, faced with no other choice than to leave their homelands as a result of catastrophic weather events, risk their lives to seek opportunities for survival and find themselves repelled on sea and land to the politics of nationalism by affluent countries upholding a fortress mentality of 'us and them'. We are living in a time when governments, global financial institutions, and fossil fuel corporations are operating against the will of a global community of people demanding climate action. Again, this makes me ask myself: why write this book? The world does not need another book dealing with issues of climate change, but to give some solace I convince myself that the world might need a book on human resistance, and in this respect, the book may offer some ideas.

I have been in this position before, that is, confronting a huge topic and my ability to produce cohesive arguments and answers to questions I have set myself. My last book, *The City in Transgression: Human Mobility and Resistance in the 21st Century* (2021), explored how cities can accommodate global human mobility as a result of displacement and destitution. The book explored the unacknowledged, neglected, and ill-defined spaces of the built environment and their transition into places of resistance and residence by refugees, asylum seekers, migrants, the homeless, and the disadvantaged. It proposed that cities should be flexible, able to swell and deflate in response to mass human mobility without repression or discrimination. Another book, *The City in Geography: Renaturing the Built Environment* (2019), reviewed the conflict between humans and physical geography from settlement to the city. The book charted the fall of geography as the unique sign of human progress and dominance over the earth's surface. It suggested that the speed of humanity's reshaping of geography for the establishment of the city displays scant regard for the earth's evolution across 4.55 billion years. It assessed the contestation of ground for the formation of the city: taming geography to meet human desires by dislocating human experience from its surroundings, by removing human experience to what is outside the city—gliding through landscapes on sealed ribbons of highways without touching the ground or flying over the earth in capsuled compartments on jet streams of compressed air in a time–space void—removing human connectivity to the natural world.

In my first book examining cities, human mobility, and geography, *Buried City, Unearthing Teufelsberg: Berlin and Its Geography of Forgetting* (2017), I explored how a people came to bury their destroyed city. The result, a human-made topographical landscape known as Teufelsberg (Devil's Mountain) where 16,000 buildings are buried, is an exemplar of forgetting Berlin's destruction during World War II. The book reviews how destruction and burial combine to reform the city in a self-anaesthesia of ruination and trauma. This book, *Sustainable Futures for Climate Adaptation, Wearing Our Ecology*, is in some ways an extension of the previous three books, but where the focus is on humanity's capacity to achieve global sustainability and restore the earth's fragmented ecological system. It covers similar themes to the previous books concerning the human condition and humanity's relation to the environment, but it departs from them by calling for a relinked human terrestrial connectivity—an ecological mobility of wearing our ecology in step with the natural world.

A previous exploration concerning human impact on the earth was through an environmental project I shared with the architect and bio-materials researcher Nancy Diniz. The project titled '*Wearing Your Air*' focused on the quality of air we breathe by designing a prototype wearable device capable of reading air particles per micrometre (PM 2.5) of pollution and displaying this information in real time.[2] Once invisible, toxic air renders air visible, becoming our unnatural enemy choking and killing millions of people with lung and blood diseases on epidemic proportions. Using PM sensors to record environmental conditions, geographic positions, and movement data, the prototype device designed as a brooch enabled wearers to record the air quality as they moved about the city in healthy to unhealthy air. The device rendered air quality measurable through LEDs displaying colour indicators to varying levels of pollution—from red to purple indicating extreme air toxicity and from green to blue for less toxic and unpolluted air. The device utilized a geographic information system (GIS) sensing pack connected to an Android-based phone running a sensor recorder to receive the data via a Bluetooth module. Google Maps was used as the base system for the GIS interface where the location is calculated by latitude and longitude values pinpointing the wearer's location as they move through the city. This data is sent to a cloud server in real time that collects and manages the data to end sensors. The data is organized through an Arduino-based control board to

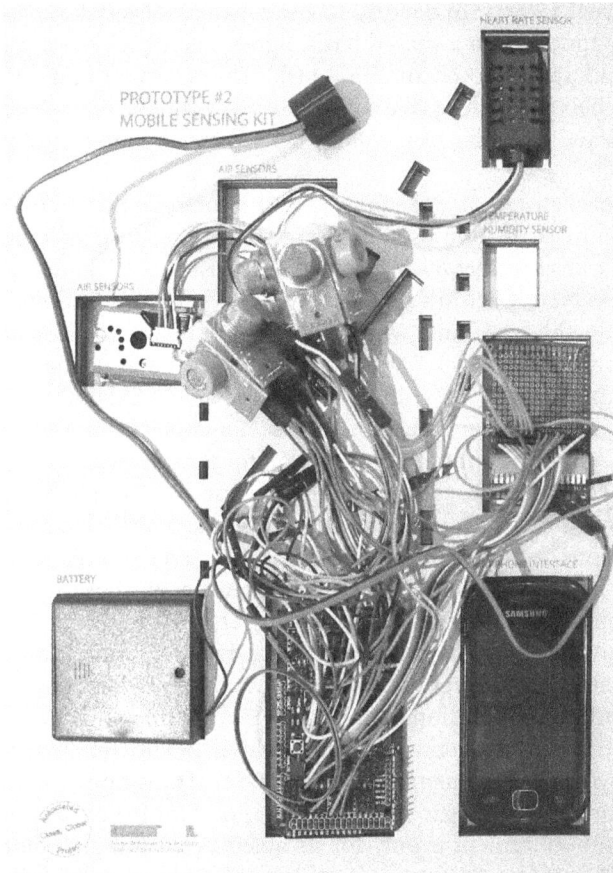

Figure 0.1 Sensing kit, *Wearing Your Air*, Lisbon Architecture Triennale.

Source: Nancy Diniz, Benedict Anderson, 2014. Exhibition image by author.

collect and coordinate the functions from sensor to phone to server. The project sought to place control of the air we breathe in the hands of individuals where device and wearer consciously initiate a new analysis of living in cities, choosing routes less harmful while encouraging a groundswell to enforce cleaner air policies.

In the face of catastrophic climate events from centuries of unchecked rampant growth, humanity is embroiled in schemes to reverse the damage it has wrought on the earth's ecologies. Global sustainability policies have, up until now, mostly focused on reducing greenhouse gas emissions by reducing fossil fuel use, halting deforestation, restoring degraded land, increasing renewable energy production from wind, solar farms, and hydrogen with the aim of reaching carbon neutrality by 2030, 2035, and 2050 depending on different countries' targets. In a world where on almost every level global exchange does not function on a shared basis, agreeing and achieving global sustainability would appear to be an uphill task. The reality of hundreds of millions of people being displaced due to the effects of climate change not of their making is a stark reminder of the divisions between peoples, countries, and continents.

Geopolitical world instability, inequality, and human conflicts consume the focus and resources of nations, relegating the urgent task of tackling climate change to a distant second place. On 2 August 2022, United States Speaker of Congress Nancy Pelosi's visit to Taiwan drew a great deal of anger from its powerful neighbour China, which in protest pulled out of its cooperation with America on shared climate change policies. Back in the United States, the Supreme Court's overriding of the Environmental Protection Agency (EPA) directive to cut greenhouse gas emissions from fossil fuel energy production plants not only paved the way for their continued use but also expanded it. Out of hundreds of examples, these two clearly highlight how sustainability is tied to geopolitics and national agendas facilitated by powerful corporations' attachment to political parties and the judiciary. The goal of achieving global sustainability is only taken up by those in power who deem it beneficial or detrimental to their agenda at the time. Another example can be given where on 8 October 2022 the UK climate minister released a press statement in response to offering 100 new licences over 900 locations for oil and gas exploration in the North Sea, claiming that it 'will be good for the environment' and 'entirely compatible' with UK climate targets. To be regulated by the government's own North Sea Transition Authority, the message that further oil and gas exploration is 'good for the environment' is an astounding announcement if only for its absurdity. With governments around the world insisting on pursuing perpetual economic growth, all that oil and gas companies have to proclaim is that they are 'maintaining energy security', which is imperative for economic security and for the re-election prospects of the political party in power.

Climate action groups such as Just Stop Oil clearly see the partnership between political parties and their corporate sponsors who place power and profit above the future health of the planet. On 4 July 2022, two members of Just Stop Oil went into the National Gallery in London and on reaching an ornate gilded framed painting, by 18th-century English artist John Constable, crossed over the rope barrier to begin their first transgression. The activists covered Constable's well-known idyllic farm scene called *The Hay Wain* with a dystopian image of air pollution, jet planes, and dead trees. The two activists then took off their jackets to reveal their white t-shirts with Just Stop Oil (JSO) painted in black across, and to finish they glued their hands to the ornate golden frame. Clear in the message of their visual stunt, the activists embodied the disparity and desperation of their action to the inaction of governments and corporations.[3] Other groups such as Extinction Rebellion,

Fridays for Future, and Last Generation support radical actions of civil disruption as a means to raise public awareness and pressure governments to act more deliberately, whereas others are criminalized for their actions.

Environmentalists of the 1960s, 1970s, and 1980s, such as the American astronomer, planetary scientist, astrophysicist, and author Carl Sagan and English ecologist James Lovelock and his Gaia theory of the earth, gave early warnings and statements. Sagan's 1985 testimonial before the US Senate Hearing on the 'Greenhouse Effect' stated that human impact on the earth was reaching a point of no return. Nearly 40 years later, climate change is only now changing how humans see themselves in relation with the earth and prompting them to implement sustainability policies to 'fix' human impact on the earth. At a time when we are witnessing individuals and organizations take on powerful industries harmful to human health, just like the civil actions waged on the tobacco industry in the 1980s, the fossil fuel companies will continue to wage war on human life and the life of the planet in exchange for profit if left unchecked. UN Secretary General Antonio Guterres recently called the fossil fuel companies' threat to human life and the earth as 'criminal' and 'suicidal'. Yet, fossil fuel companies are only one symptom of a world where capital, production, and consumption dominate, wherein an endless stream of products is destructively intrinsic to human life on the earth.

In the 21st century, humans have come to play the social role of consumers first and then another role as responsible climate change activists. Where governments flounder through inaction and fossil fuel companies continue their drive for profit, it is individuals and groups who are engaged in making the radical changes necessary to bring capital down and elevate the health of the earth for future generations. Catching on to this changing tide, companies have deployed the term 'Green Growth' by claiming that their products are sustainably produced as a way to reinforce that it is beneficial to keep on consuming their products—a policy known as 'Greenwashing'. Counter to these terms is the notion of 'Degrowth' that points consumers in the opposite direction, asking what they really need for their day-to-day lives. In other words, the message is to think before buying and to make do with less. These terms, lies, and malpractices mostly involving rich Western nations pale by comparison with the terms used to demarcate global divisions, most notably the 'Global North' and 'Global South'. Mispronounced, misplaced, and geographically dislocated, the terms are a timeless piece of propaganda peddled by the Global North countries in suggesting that they assist the development of the Global South. Yet, in reality it is the other way around. Exploiting the resources and cheap labour, it becomes clear that the Global South maintains the wealth of the Global North. Poor countries are then developing rich countries and this is reinforced by the imbalance of the economic exchange. For every US$1 the Global South makes, the Global North makes US$30. This 1:30 exchange rate is based on resource extraction and labour to the profits of mineral processing, product production, and sale.

We are reaching what is being termed *Earth Overshoot Day*. As with the term 'peak oil' where global production reaches maximum output to global consumption, we are now consuming 1.75× more resources than are being replenished. Earth Overshoot Day is calculated 'when humanity's demand for ecological resources and services in a given year exceeds what Earth can regenerate in that year. We maintain this deficit by liquidating stocks of ecological resources and accumulating waste, primarily carbon dioxide in the atmosphere...Earth's Biocapacity/Humanity's Ecological Footprint × 365 = Earth Overshoot Day'. In 2022, Earth Overshoot Day was reached on 28 July where humanity 'used all the biological resources that Earth regenerates during the entire year'.[4] Most of the appliances and devices we buy are programmed with in-built obsolescence to reduce their lifespan so that the new and latest appliance and device will be brought. On the internet, YouTube and Instagram influencers vie with each other to promote and exploit brands of fast fashion and devices where each 'view' and 'like' increases their earning capacity. Swiping and tapping on your smartphone, the selection and purchase of an item is completed within seconds and in no time the item ordered is shipped from a different country thousands of kilometres away to arrive on the doorstep replacing the old item that has gone out with the trash. How can sustainability be achieved when consumption has become so deeply attached to human 'satisfaction' and 'staged' to governments and industrialists in their pursuit of endless growth? The answer is it can't. One proposition advocated by progressive economists is to rethink the labour market. The idea is to make provisions for people to work less by sharing jobs so that extra people have time to invest their labour into ecology programmes that benefit their environment, region, country, and the planet. To conservative economists such a proposition is naïve, to companies it is economically unviable, and to governments it is a big problem where citizens organize alternative societal structures and take control of their lives with less administrative interference.

To avert environmental collapse requires global cooperation. The 27 COP conferences held so far, alongside individual countries' efforts to limit fossil fuel use, reduce CO_2 emissions, and maximize sustainable energy production, have no doubt become more consistent in their programming across the world. Yet, when world leaders are concerned with holding on to power, maintaining global capital inequity, pursue economic growth, increase military spending, increase nuclear weaponry, restrict gender equality, institutionalize racial discrimination, and contain freedom of movement, how can global sustainability be achieved under these agendas? In no uncertain terms, it requires a revolution on a scale not undertaken by humanity in order to change the trajectory that has so far determined human relationships with the natural environment. Humanity has a myriad of revolutionary histories and these are not just the slave, peasant, political, and religious revolts by usurpers to rulers of the day. It is the larger revolution of human behaviour that is required. In the transition from nomadic life to settlement 12,000 years ago to the industrial revolution and the mass excavation of the earth's resources, production and consumption of goods in the 19th and 20th centuries, humanity has asserted its control over the natural world. For humanity to disentangle its domination over the earth, a defining task for life in the 21st century is to recalibrate and relearn how to inhabit the earth.

How humanity relinks with the earth is dependent on how it reimagines its previous existence on the earth. A part of the reimagining is to redefine humanity's physical manifestation with the natural world; the ecology of animals, plants, and all living things to unearth the 'nature' of human nature. This is something explored in the 2022 Venice Architecture Biennale's The Milk of Dreams curated by Cecilia Alemani. Alemani's Biennale themes are based on how humanity intends to proceed into the future with the challenges of climate change, sustainability, and the looming extinction of animal and plant species. 'How is the definition of the human changing? What constitutes life, and what differentiates plant and animal, human and non-human? What are our responsibilities towards the planet, other people, and other life forms? And what would life look like without us?' These are just some of the questions Alemani sought to pose, as well as asking what adjustments humans can make in response. These are big questions and Alemani doesn't stop there, proposing 'the representation of bodies and their metamorphoses; the relationship between individuals and technologies; the connection between bodies and the Earth'.[5] Chapter 4, Weathering Patterns, explores animals and plant life transmutation and metamorphosis and how humans might learn, adopt, and adapt for their transformation to living on the earth.

Creating new codes of behavioural presence to redefine our physical impact on the earth requires building new language pathways to communicate with our ecologies. All existing languages within nature offer a base for humans to unearth their own nature. Each interpretation of this new language and presence in the natural world is reconnecting ecology, weather, geography, terrain, ocean, river, plant, animal, fish with our own. The lexicon that emerges is geographically located, ecologically rooted, and physically experienced, a language to evolve humanity to a new course in ecological coexistence. So vast is the idea for ecological coexistence between humans and the earth as to be overwhelming to achieve. 'We want coexistence to mean the end of narcissism, but this is an agrilogistic thought that would destroy in advance the relation to the other' writes Timothy Morton in *Dark Ecology, For a Logic of Future Coexistence*. 'It is difficult to think agrilogistically in the face of our emerging awareness that we are a hyperobject (species) inhabiting another hyperobject (planet Earth)'.[6] Ecological coexistence could be understood as a new form of environmental prospecting where humans allow nature to prospect them just as they have prospected nature. In terms of practice, how this new prospecting comes about is to review regions at risk due to deforestation, desertification, drought, extreme heat, and the threatened extinction of animal and plant species and pursue a programme of restoration financed by surplus producing regions. There are growing calls for financial reparations from former colonial countries to their colonial invaders; and at the 2022 COP27 conference, agreements and financial commitments between countries were made. As the future of human and Earth coexistence is dependent on constructing a new connection to the earth, confronting histories of invasion and the foundations of global inequality is a prerequisite to achieving global sustainability.

International financial organizations such as the UN and the World Bank are engaged with dramatically shifting world inequality by fighting global food insecurity, addressing clean water, redirecting excess food production, limiting waste, and supporting new forms of energy production to replace the burning of fossil fuels. The amount of adjustment the world needs to undertake is as immense as the scale of environmental devastation it has wrought. Addressing global wealth inequity challenges centuries of European/Western global domination and faces powerful opposing forces, a lack of political will, and many forms of corruption. Under such conditions it could be assumed that reaching global sustainability is a fantasy given the global economic structure, protectionisms, and greed that divide the world between rich and poor, developed and developing nations. Yet, the only way out of the present crisis is a direct assault on the global status quo for the redirection of wealth. At a basic level of understanding how sustainability might actually work, it is necessary to first acknowledge that two billion people live with food insecurity, basic or no sanitation, and limited access to clean water. It is to also recognize that indigenous peoples across the world are daily threatened and forced from their homelands due to industrial-scale plantation farming, illegal logging, and cattle ranchers. As rich Western countries find ways to fix their CO_2 emissions output, a far greater majority of the world's population struggle to meet their daily needs for survival. It is clear that disruptive human impact on the earth requires more than financing sustainable practices; it needs the wholesale restructuring of humanity. Ideologically utopian and naïve, such restructuring is at odds with a world driven by divisions, profits, threats, and fears. But what are the alternatives to averting further irreversible environmental damage and the forced migration of hundreds of millions of displaced people? If humanity is unable to make the necessary adjustments to the catastrophic turbulence the earth now faces, then predictions of humans and the earth ceasing to exist as we know them are not overstated.

What if humans were to regain their nomadic ancestry to become extraterrestrials? Enlightened humans able to advance their 'nature' to live biologically connected to animals, plants, and all living things? To do this would require a rewiring of the human psyche and its long-held position of being at the centre of the world. To achieve a sustainable human and Earth coexistence, human intelligence needs to re-engage and cross the boundaries that have divided it from the natural world. Living sustainably requires a lived biology and a whole-Earth–human ecology. Humans must rid their prospecting bodies of gross consumption, forge equity, overturn capital, and build systems focused on humanity's relations with the earth, for then the terrestrially mobile human world stands a good chance of forging human and Earth future. These terrestrial future humans will take generations to evolve and wear their ecology as a natural course of being in the world. But this seems a far better proposition than wearing out the earth.

The relation between humans and the earth has always been uneasy but it has never had so much disruption as it does now. The problem was always one of scale. The scale of the earth and all living things versus the pared-down scale of humans with all their intelligence. The sedentary life of the settlement brought an end to nomadic life and never did it occur to the original settler that through their actions they were building-out the ecology from their lives in exchange for guaranteeing their survival. As timidity and fear encased the settlement to what laid outside its walls, a new perception formed that nature would need to be tamed, controlled, and placed in service of the desire of human enterprise. This story is only partially true, but it was enough for the story to flourish and spread across the world. On the scale of the earth's existence, this story may be but a speck, but it is an incredibly destructive and growing stain. Humanity is now faced with its most perplexing problem: what to do with itself?

The structure of the book explores various perspectives in understanding and countering climate change. Each chapter is divided into three sections addressing various aspects of climate change, sustainability, and humanity. Chapter 1, Sustainability's Paradox—Commitments and Inactions, charts various conflicts affecting global sustainability as a working model for reducing human impact on the earth's ecologies. The first section titled Fraud—Environmental History gives a brief account of the conflicts associated with sustainability and how it is subjected to and constantly redefined and repurposed by governments and corporations to suit their agenda. The section illustrates the various ways in which environmental global commitments reached between countries often fail in the light of the power struggles of geopolitics and capital. The second section, Climate Adjustment—Combating Global Inequality, reviews how the terms such as 'Global North' and 'Global South' are used to describe global economic division rather than addressing the histories of colonialism and present-day corporate imperialism. The third section, Action—Climate Resilience and Decarbonisation, gives an overview of the present practices in reversing carbon emissions. The section gives various accounts of governments, institutions, and the recent COP27 conference in Sharm El Sheikh in relation to the issues of climate change and sustainability and what people around the world and organizations are doing to obtain climate change resilience.

Chapter 2, Terrestrial Migrations—Nomadic Ecologies, charts early human nomadic migration to the formation of settlement and the present-day digitally accessible world. Roaming—Earthy Freedoms reviews how human migration out of Africa established human ecological culture across the earth through a gathering of climatic knowledges and geographical adaptations. It suggests that human relatedness with the natural environment was radically altered by the static occupation of ground. The second section, Settling—Restraining Fields, reviews how fear and timidity seeped into human relations with the natural environment through the contestation with nature. What emerged from an imaginary threat from the natural environment created a human psychogeography condition—a borderline disorder where humans fragmented their natural ecology. The third section, Territorial—Digital Accessibility, builds the progression of human separation from the natural world to the separation of regions in the creation of nation state territorial boundaries in opposition to the digitally accessible world of today of global communications. The fear and timidity of the early stages of settlement to the militarization of borders and global deterrence of refugees is further dividing the world in the face of the non-territorial boundaries of climate change.

Chapter 3, Earth Extractions—Pillage and Ransack, takes a broad approach by focusing on what happens when global separations on the basis of class, race, gender, and opportunity are torn down, protectionism removed, and economies and societies reformulated. Plunder—Exploitation and Destruction reviews the exportation of the European denaturing of the world beginning with the Age of Exploration, European invasion and colonialization, most notably on the African continent and the suppression and enslavement of its original occupants spurred by the racial discrimination and greed of Western industrialization. The second section, Estrangement— Antagonism, Disaffection, Hostility, asks how the separation between urban and rural populations' exposure to climate change can be reconciled through a shared cooperation to close the gap. Urban dwellers' experience of climate change comes via estrangement by constructing surfaces over ground; establishing a veneer of sealed spaces to maintain physical control through the erasure of natural ground. Rural populations on the other hand experience estrangement to their environment as a result of weather turbulence directly affecting their survival. To formulate a strategy for a sustainable future world requires folding-in these two distinct experiences through a far greater human exchange. The third section, Equality—The Matter of Sustainability, asks if the world is ready, especially rich Western nations, to give up their privileges of rampant consumption to keep servicing their lifestyles for a redistribution of wealth between rich and poor nations.

Chapter 4, Weathering Patterns—Entering the Biosphere, charts how connections with weather have become more acute and mediatized. Brought about by unpredictable meteorological turbulence due to climate change, the impact of weather is now more than ever connected to human survival. Weather is intrinsic to all living organisms, yet humanity is obsessed with building-out weather. Early humans' use of the geological stone to advance civilization evolved to characterize an inner and outer 'geophilia' to the natural world. Modern-day 'geophilia' has created a 'pictorial' view of weather, reducing human environmental comprehension and adaptation prevalent in the transmutation and metamorphoses of animal and plant life. The first section, Animal—Ecological Comprehension, looks at the distinction of humans and animals and how the former understand themselves in ecology. The section looks at animal and human relations 'across the narrow abyss of non-comprehension' might be bridged, where humans re-establish their relationship with the world's ecologies having abandoned them in favour of technological progression. The second section, Plant—Society in Botany, reveals the complexity of plant life and takes a similar view concerning human relations to plants 'by which nature produces one part through another, creating a great variety of forms through the modification of a single organ'. The metamorphosis of plant life has far outstripped humanity's evolutionary cycle and the section takes a view on how humans can build a mutual transformation with the ecologies of plant life. The third section, Human—Animal–Plant–Weather, brings the previous two sections together in how humans might comprehend the natural environment as not separable from themselves. It asks what steps are integral for 'human metamorphosis', an enhanced human ecology and Earth sustainability alongside animal and plant life.

Chapter 5, Climate Gathering—Wearing Our Ecology, applies the concept whereby ground moves as humans move over it, creating an intrinsic cartography of the land. Exploring the concept of 'wearing our ecology' where human mobility is synced and merges with landscapes, topographies, and geographies, the chapter threads conceptions for human–ecology–Earth coexistence. The first section, Border Lines—Climate Cartographies, reviews how human ecology was dislocated through cartography by representing terrain in spatial terms through lines and marks rather than physical characteristics. It became the defining spatio-temporal experience to map demarcating regions into geographically unrelated territories ceasing at invisible borders. The second section, Climate Moves—Migrating Sustainability, looks at how experience and knowledge of terrain collected in early nomadic life fundamental for their survival underwent a transformation as human invention, technology, and building, most notably in the 20th century, undid humanity's historical footing, resulting in lost ecological knowledge for adaptation. To rebuild connectivity with ground that is instrumental for human and Earth sustainability allows new forms of cohabitation to emerge; chances arise, space extends, and the territories of the mind magnify. The third section, Mobile Ecologies—Environmental Costuming, asks what it 'means' or 'looks like' to 'wear our ecology'. The present run on finding technological solutions to reduce human impact on the earth, such as capturing greenhouse gas emissions from industry, methane gasses from farmed animals to creating atmospheric discs to deflect solar rays from the sun to decrease global warming, are trundled out almost on a weekly basis. 'Wearing our ecology' begins with planning new living strategies, rethinking the financial systems of capital, industry, and consumerism that have generated unlimited gratification at the expense of the earth.

Chapter 6, Environmental Adaptations—Spiral-Swarming, Human Diversity, gathers the ideas and examples from the previous chapters to examine adaptable solutions for future human and Earth coexistence. One of the aims of the chapter is to better understand the concept of a wearable ecology for adaptive living by extolling a mobile human-geography in perpetual cycles of exchange, gathering, replacing, and recharging natural environments. The first section, Oscillations—The Climate Spiral, asks how a global pact to encourage sustainability can be fully implemented given a great percentage of the earth's pollution is generated by rich Western nations and their fossil fuel industries such as the United States, Europe, Russia, the United Kingdom, Saudi Arabia, Qatar, and Australia and developing nations such as China, India, and Indonesia in contrast to the world's poorest regions, specifically Africa, South America, and Oceania which experience first-hand the effects of climate change. Taking the concept of swarming that allows for new human patterns of habitation to emerge, the section addresses how a global agreement to reduce climate change can be implemented when an acute imbalance causes sustainability to spiral in two opposing directions. The second section, Reforming—Living-with, Building-out Nature, considers how to address the division of embodied nature between the approximately four billion people living in dense urban centres in contrast to the world's other four billion people living in rural areas. Differentiated through (dis)connections to the elements, Sun, rain, heat affecting the cultivation of crops to food on supermarket shelves, illustrates the physical/visual divide and fundamental stumbling block to achieving global sustainability. The third section, Variations—Biological Departures, explores how humanity can depart from its historical ascendency and superiority over nature as a necessary step to achieve a sustainable future world. The persistence of human nature in the belief that it can restore the earth by re-engineering the same technologies that destroyed it indicates a perverse account of humanity's psychological attitude. Developed over thousands of years of shaping the environment to fulfil human desire, it is now imperative to reverse this attitude. The section asks how can embodied nature be resurrected as an essential element for individualized and societal sustainability?

Chapter 7, Future Human—Ultra-Terrestrial Worlds, begins by presenting human and Earth coexistence through the lens of science fiction. Our understanding of the future comes through books and films that vividly imagine a future world in mostly apocalyptic scenarios, which would seem to indicate that humans are preparing for their own destruction. Simultaneously harrowing and visually entertaining, these fictive futures appear not to shock but to connect the present realities of conflict and adversity across the world and make them strangely compatible. Films and series such as *The Road* (2009), *The Book of Eli* (2010), *Contagion* (2011), *The Colony* (2013), *World War Z* (2013), *Resident Evil* (2021), *Dune* (2021, 2023), *After the Pandemic* (2022), and *Black Knight* (2023) visualize the earth in decline, where the natural world is destroyed, despotic leaders rage, and space is lawless and humanity degenerate. Set in 2080, *Future Human* is a film pitch to studio executives that deploys and reiterates the ongoing narrative of a dystopian future world and the epic struggle for the earth's future in this case between the 'Technophiliacs', who believe that the world can be saved through human innovation, and the 'Neoecologists', who believe that a complete transformation of human relations with the natural world is needed. Anarchic activism clashes with hierarchical structures of corrupt geopolitics and governments, religious suppression, and inequality. Militarized territories are pitted against human mobility, and indigenous-led environmental solutions take hold to radically shifting climatic forces. Can the longstanding conflicts between the technologists and ecologists be resolved through human intelligence or are the forces of violence, corruption, and control what drive human life on Earth closer to extinction? The chapter finishes with what might be interpreted as a more serious 'academic' summary of the ideas contained in the book. Appropriately titled Epilogue—Factual Reality, the section pulls together the concepts explored in the book to offer a set of outcomes as much as a critique of how the book has tackled the vast and complex subject of climate change, sustainability, and the future of human and Earth coexistence. The summary raises questions rather than delivering concrete answers or providing solutions, much of which is beyond the ability of any single person and this author.

The ideas and explanations contained in this book will not affect how humans live on Earth, how governments and institutions increase their responsibility, and fossil fuel companies and global mining corporations reduce their impact on the climate. Humanity's relations with the natural environment from nomadic life to the establishment of settlements to the industrial age and on, to the globalization of today, have been bent into now unrecognizable forms. The idea of presenting a future human society based on terrestrial mobility and climate adaptation for 'wearing our ecology' means unearthing the buried terrestrial 'nature' of human nature in securing a livable world for future generations. The immense amount of information generated daily concerning climate change events and sustainability policies has been overwhelming and impossible to keep up with. Nevertheless, I have cited a substantial amount of sources from various UN climate change conferences, NGO programmes, climate activist campaigns, and commercial projects from released reports and media coverage. But where these move and shift, go forward, and step backward in tackling climate change, the human dimension remains consistent: inequity between rich and poor, histories of colonialism, racial discrimination, resource theft, religious intolerance, disproportionate gender disparity—all of which affect the real chances of reducing human impact on the earth. Addressing these inconsistencies is fundamental to achieving global sustainability. Other references are sourced across a board spectrum of books on global warming, climate change, sustainability, adaptation, ecology, the biosphere, environmental science, meteorology, urbanism, rural-regional, migration, human geography, colonialism, capitalism, and corporate imperialism, as well as works on botany, plants, and animals. I consider sustainability to involve understanding humanity's past relations in finding a way forward to reform humanity's psychological dislocation that has shaped its interaction with the natural world. To that end, this book offers a way of thinking by offering concepts and directions for 'wearing our ecology' in the age of climatic turbulence.

Notes

1 For the full text of UN Secretary General Antonio Guterres' speech, see www.yout ube.com/watch?v=k0bskMLyMcA
2 For more information concerning the project and collaboration with designer and architect Nancy Diniz, see article 'Wearing Your Air' in *Unconventional Computing: Design Methods for Adaptive Architecture*, edited by Rachel Armstrong and Simone Ferracina. Ontario: ACADIA and Riverside Architectural Press, 2013, pp. 172–175.
3 In response to their action, a JSO spokesperson said they had attached an 'apocalyptic vision of the future' that depicts 'the climate collapse and what it will do to this landscape'. One of the activists declared the reason for their intervention: 'When there is no water, what use is art? When billions of people are in pain and suffering, what use then is art?' Of course, art is not the problem here except for the privilege where art resides in culture and value over the lives of tens of millions experiencing extreme poverty and life insecurity due to climate change. What the activists are doing is to challenge our values and our moral fortitude in an era of globalization and telecommunications. John Constable's painting is not a guilty party here, but anyone looking at it, the institution protecting it, and the culture that cherishes the nostalgia of rural 19th-century England is not focusing on the real and present danger of inequality, poverty, and devastation wrought by climate change on hundreds of millions of people throughout the world. A few months later on 14 October 2022, two other members of JSO threw a can of tomatoes on the glass security pane of Van Gogh's famous painting Sunflowers, also at the National Gallery. Once the soup was wiped from the glass and the painting put back on the wall, anyone paying attention might not just see a bunch of yellow sunflowers when looking at the painting but dripping red tomato soup; a symbiotic blood–oil image of today's world and British Petroleum's sponsorship of the British Museum at the time. For more information on JSO's intervention at the National Gallery, see www.independent.co.uk/climate-change/climate-protesters-glue-consta ble-b2115665.html
4 For more information on Earth Overshoot and also the Global Footprint Network, a research institute which calculates the 'Earth's Biocapacity/Humanity's Ecological Footprint', see www.overshootday.org/
5 For more information on the 2022 Venice Architecture Biennale, see www.labienn ale.org/en/news/biennale-arte-2022-milk-dreams
6 See Timothy Morton, *Dark Ecology, For a Logic of Future Coexistence*. New York: Columbia University Press, 2016, p. 105.

Bibliography

Diniz, Nancy, and Benedict Anderson. 'Wearing Your Air'. In *Unconventional Computing: Design Methods for Adaptive Architecture*, edited by Rachel Armstrong and Simone Ferracina. Ontario: ACADIA and Riverside Architectural Press, 2013, pp. 172–175.

Morton, Timothy. *Dark Ecology, For a Logic of Future Coexistence*. New York: Columbia University Press, 2016.

1 Sustainability's Paradox
Commitments and Inactions

Fraud—Environmental History

The power of human beings to affect and control and change the environment is growing as our technology grows, and at the present time, we clearly have reached the stage where we are capable, both intentionally and inadvertently, to make significant changes in the global climate and in the global ecosystem... Because the effects occupy more than a human generation, there is a tendency to say that they are not our problem. Of course, then they are nobody's problem; not on my tour of duty, not on my term of office. It's something for the next century; let the next century worry about it. But the problem is that there are effects, and the greenhouse effect is one of them, which have long time consequences. If you don't worry about it now, it's too late later on. And so, in this issue, as in so many other issues, we're passing on extremely grave problems for our children when the time to solve the problems, if they can be solved at all, is now... I think that what is essential for this problem is a global consciousness view that transcends our exclusive identifications with the generational and political groupings into which by accident we have been born. The solution to these problems requires a perspective that embraces the planet and the future. Because we are all in this greenhouse together.
(Excerpt from Carl Sagan's testimonial before the US Congressional Committee on the Greenhouse Effect, 10 December 1985)[1]

American astronomer, planetary scientist, astrophysicist, and author Carl Sagan's testimonial before the US Senate Hearing on the Greenhouse Effect on 10 December 1985 revealed the nascent awareness of the scientific and political elite of the consequences of climate change at the time. Sagan's 20-minute talk on the greenhouse effect (as it was then referred to) is not only a straightforward scientific report on what greenhouse gases in the earth's atmosphere mean for the health of the planet but also a philosophical, ethical, and moral accountancy concerning the lives of future generations. Connecting the built-up greenhouse gases in the atmosphere to the rise in the earth's surface temperature as a result of a loss of the reflective heat

DOI: 10.4324/9781003382515-2

shield protection from the sun is causing catastrophic consequences such as rising sea levels, land degradation, drought, flooding, and melting ice. Listening to Sagan's talk is the young US Democratic Senator Al Gore who understood the need for greater research and knowledge on the greenhouse effect and who, following his presidential election loss in 2000, became a leading climate advocate, most notably in his ground-breaking documentary film *An Inconvenient Truth*.[2] A decade before his testimonial, Sagan was commissioned by NASA to design a 'welcoming plaque' stuck to the outer layer of its Laser Geodynamics Satellite (LAGEOS), a spacecraft designed to measure continental drift. Consisting of three graphics, each depicting the formation of the earth's continents in past, present, and future time, the 'welcoming plaque' also contained a golden record disc of greetings in 55 languages, 115 images of people and the earth, an eclectic selection of music including a Bach concerto and Chuck Berry's 'Johnny B. Goode' and sounds of surf, wind and thunder, birds, whales, and other animals. The plaque and interstellar messages, images, and sounds were aimed at extra-terrestrial life to play on their record players as much the earth's future inhabitants in the year 8,000,000 as a way of explaining previous human existence on Earth just in case all historical records had been erased. Shot off into the cosmos on 4 May 1976, *Voyager* is still orbiting its pathway of discovery in the universe; a drifting relic of scientific antiquity and warning.[3]

During the 1970s, environmental scientists were primarily concerned with the earth's ozone layer, specifically the ozone hole over the Antarctic and 'acid rain' in the atmosphere as a result of sulphur dioxide from the coal-fired power plants, industry, and leaded petrol. For every ton of coal mined and smelted, 10 tons of waste, primarily CO_2 and coal ash containing mercury, arsenic, and cadmium, is produced, which contaminate ground water and river systems. These polluting gases mixed with the cloud formations oscillating the Earth and rained in a toxic deluge known as 'acid rain' onto the ground. After decades of neglect, environmental protection laws were passed in many countries and a trading system of excess carbon emissions set up to cap sulphur dioxide in the atmosphere. A two-pronged approach was created: one aimed at removing the threat of acid rain; the other at repairing the ozone layer, specifically the ozone hole over the Antarctic. Accompanying these actions was the banning of the use of chlorofluorocarbons (CFC) in refrigeration, aerosol cans, and solvents, and perhaps most importantly the phasing-out of leaded petrol and phasing-in of the use of unleaded and ethanol fuels and the installation of catalytic converters in cars. All of these measures were a result of a concerted global effort between countries that amounted to a successful campaign. While limits, bans, and devices reduced the amount of sulphur dioxide in the atmosphere, and led to the substantial repair of the ozone layer, it did not stop the use of fossil fuels. Their continued use has increased and subsequently we see the exponential rise in global carbon dioxide emissions in the atmosphere resulting in global warming and catastrophic weather events across the world today.

Attempts at creating global sustainability policies have been ongoing since the first United Nations sponsored Conference of the Parties (COP). Made up of countries who signed up to the United Nations Framework Convention on Climate Change (UNFCCC), the first conference was held in the German city of Bonn in 1995 and since then it has been held annually in various countries across the world. At the time of writing, there have been 27 COP conferences where agreements on CO_2 reduction targets have been set, pledges to renewable energy production made, and commitments to partial carbon neutrality by 2030 and fully by 2050 promised. But what has been agreed on paper does not necessarily transfer into reality. Since 1995, oil and gas consumption has increased, with renewable energy uptake consistent and increasing though still inferior overall to fossil fuel energy production. Deforestation has continued to rise as has land and sea degradation, extreme weather cycles, increases in food insecurity, and water scarcity affecting billions of people around the world. All these effects on the earth and human life we have come to know as being attributed to climate change. Yet, these dire consequences are part of a larger interconnected problem. The disparity in economic inequality between rich and poor nations and gender inequality in patriarchal cultures where women and girls mostly responsible for providing food and gathering water significantly suffer more from climate change and ineffectual sustainable policies. Human conflict has also increased due to climate change where various militant factions, herders, and crop farmers fight over resources, land, and water. Not directly related to climate change but to the decisions of despotic autocratic power, civil war and religious terrorism can and do effectively derail a region and country's sustainability programmes. The 2022 Russian invasion of Ukraine has exposed the fragility of global food resources, particularly in countries dependent on Ukrainian wheat, corn, and sunflower oil. Middle Eastern and African countries heavily dependent on these products have faced severe food shortages further adding to the global population in excess of two billion people experiencing hunger on a daily basis according to a UN report.[4] Russia has also spread its invasion by weaponizing its oil and gas supplies to Europe to drastically cut capacity and create an energy crisis and disrupting European carbon reduction targets. While the cut in Russian oil and gas supplies to Europe has resulted in a greater focus on renewable energy, it has also extended the life of coal and nuclear power stations to meet the shortfall. In the unfolding crisis, the European Parliament declared both sources of energy production as 'sustainable', legitimizing the practice of 'greenwashing' and further jeopardizing the European Union's ability to reach carbon neutrality by 2050.[5]

On the other side of the globe in the United States, the Supreme Court on 31 June 2022 handed down its 6–3 ruling against the Environmental Protection Agency's (EPA) statutory powers to limit greenhouse gases emissions. The judges voted along conservative/liberal lines to restrict the operations of coal-fired electricity plants in mostly Republican-held states as unlawful. The ruling sets back not only the Biden administration's Clean Air Act policy but also its international obligations for carbon emissions reduction as agreed in the 2015 Paris Climate Summit. The Supreme Court's decision is startling for it reveals how bipartisan politics can overrule not only international agreements but also the majority of America's population, which are committed to reduce carbon emissions, in a shocking indictment of a government body sworn to be politically neutral. The Supreme Court's ability to remain politically neutral and religiously nonaligned was further compromised in its reversal on 24 June 2022 of the 1973 Roe v Wade decision (giving the right of women to choose abortion), thereby meeting one of the demands of the Christian right, which is indicative of the fossil fuel, energy, and mining companies' ability to exert their agenda over these lawmakers. Given America's politically polarized system, climate change action in the country was given a massive boost on 8 August 2022, when the US Senate passed 51–50 along Democrat and Republican party lines with Congress following suit, the Biden administration's $369 billion Climate Action Bill to reduce carbon emissions by up to 40% by 2030. Yet, to secure full Democratic senate support where no Republican senators voted for the bill, a very important cache was included to guarantee the protection and continued use of fossil fuel energy production in key states. What this confused message tells us is that the four layers of US governance—White House, Supreme Court, Senate, and Congress—have meshed political ideologies with climate change action as distinctly manipulable and voluntary; fracturing its ability to realize sustainable policies.

Amid success, disunity, and setbacks, achieving global sustainability is not limited to European or United States conflicts, politics and carbon emissions targets; it is pervasive throughout the world. Major fossil fuel suppliers such as Russia, Saudi Arabia, Iraq, Iran, and Australia and major manufacturers and polluters such as the United States, European Union, China, and India are aligned to pursue continued economic growth first and environmental restoration second. Atmospheric weather turbulence resulting in land degradation, desertification, acidification, floods, droughts, extreme heat, deforestation, fresh water scarcity, oceans of plastic, fish stock depletion, and animal and plant species extinction speak of a failure to respond to the emergency of climate change. Added to these failures is government and corporate corruption, wars, and terrorism while hundreds of millions of displaced people have had to abandon their homelands through no choice of their own and either reside in refugee camps or risk their lives to reach Western countries in the hope of having a better chance for survival. What may appear as a series of insurmountable obstacles simply reflects the viability of reaching global sustainability. Around the world, short-term economic growth still outweighs long-term sustainability goals. This can be highlighted by Japan's Prime Minister Fumio Kishida's statement in June 2022: 'I want to achieve a virtuous economic cycle by raising the incomes of not just a certain segment, but a broader range of people to trigger consumption. I believe that's the key to how the new form of capitalism is going to be different from the past'.[6] Prime Minster Kishida makes clear that economic growth and consumption remain at the top of his government's agenda and any mention of sustainability is noticeably absent. Such mixed messages concerning sustainability and the economy and political polarization have muddied the waters and sown public division and it is increasingly falling on the will of the people to demand climate action.

In *Hijacking Sustainability*, Adrian Parr explains that the burgeoning culture for sustainability works out in two major divisions: society as a whole and the climate change activists within it, and the greenwashing fossil fuel companies which deploy the notion of sustainability to rebrand themselves as socially responsible corporations. Parr argues that developing sustainable practices 'meets the needs of today without compromising the needs of future generations' where 'the movement seeks justice for the underprivileged, including the right of the environment not to be destroyed'.[7] Petroleum, mining, manufacturing, transport, industrial food producers, and supermarkets, whose financial operations are intertwined and who contribute a major percentage of CO_2 emissions, have become major exponents at marketing their environmental policies while continuing to rely on fossil fuel and non-renewable energy sources for their operations. For environmental protesters these global companies' hypocrisy and co-opting of the social message amounts to institutionally sanctioned criminality. Known as 'eco-branding' and 'greenwashing', these companies are adept at getting their message across through organized media campaigns while seeking new avenues to expand their oil fields, destroy rainforests through mining operations, continue to manufacture limited shelf-life products, pursue water-intensive food production, and plastic packaging—all of which led to global carbon emissions reaching their highest level ever recorded in 2022. According to Parr, 'the greenwashing thesis presupposes that the greenwash misrepresents the fundamental principles of sustainability culture. Undeniably, in its focus on the coercive power of branding, the theory of greenwashing ignores the affective power of ecobranding'.[8] What eco-branding and greenwashing does is allow fossil fuel corporations and associate industries to modify their operations by putting forward a mediatized model of environmental responsibility while protecting their industry and vast profits at the expense of the planet. Capturing, marketing, and hijacking the sustainability message is easy picking for these powerful actors, making it '*meaningless* if the ecosystems and biodiversity needed to maintain the quality of life of future generations are not supported and cared for'; moreover, 'they do not fully address the fact that sustainability is also *useless* if we reduce or restrict the dynamic and creative energy of life to profit-maximizing economic principles'.[9] If we are to believe that sustainability is '*meaningless*' and '*useless*' as Parr suggests, what are the alternatives? The reality is sustainability and climate change policies are the enemies of fossil fuel corporations just as inequality is the friend of global capitalism and economic division.

Sustainability remains disempowered in the face of the forces of absolute power and capital as it does to racial, financial, and gender inequality as it does to air and plastic pollution. Where political opportunism, corporate profiteering, and corruption control the extraction, processing, and burning of the world's natural resources, and where patriarchy discriminates and reduces the abilities of women and girls in society, sustainability suffers. As term, ideology and practice, sustainability is aligned with global economic, racial, and gender equity. In terms of the whole-Earth organism—human, plant, animal, and geology—sustainability must address humanity's exclusive control over the earth and all living things. Without seriously tackling these issues, sustainability will remain unachievable. To acquire global sustainability it necessitates a rethinking of humankind's position to all living things on the earth. Reimagining humanity's long-held illusion as the supreme entity on the earth requires a total dismantling of its psyche of dominance over the natural world.

To contend with the world's radically shifting and unpredictable climate, sustainability has become another piece of human invention to be commoditized like any other. It sits at the end of the line of humanity's progression as a way forward to reduce its impact on the earth. While sustainability sits on the good side in the fight against climate change, on the other side are fossil fuel companies which remain emboldened to continue to risk the world for profit with the support of governments. Seen as saving the earth from catastrophic destruction, the broad agenda of sustainability is plundered for propaganda purposes by governments and corporations appearing alongside the people as street fighters tackling climate change, meaning that the climate emergency is constantly being undermined and overwritten. When there is an energy crisis as illustrated by the Russian war in Ukraine, fossil fuel companies validate their ongoing extraction and infrastructure investments as strategic national interests. Norway's announcement of investing in new oil and gas sea exploration as a necessity to advert energy shortage within Europe and the Democratic Republic of the Congo's (DRC) auctioning of 30 licences to mining and energy companies opening vast swathes of the Congo Basin to exploration, endangering the world's second largest tropical rainforest's ability to absorb CO_2 emissions, are examples of national and corporate interests taking precedence over a global emergency.

But who is to blame here? In the case of the DRC's opening of the Congo Basin to exploration, its government and industry partners argue that it will bring benefits to one of the poorest nations on Earth. Will it? A glance across the world in terms of individual CO_2 emissions illustrates the disparity versus the responsibility. For example, in the United States, individual CO_2 emissions average at 16.2 tons per year, in the European Union 6.8 tons, in the Middle East it reaches 49 tons per year (e.g. Qatar), and in Sub-Sahara Africa the average comes in at 0.8 tons per year.[10] Across the whole of the African continent, CO_2 emissions amount to 3.7% of gross CO_2 global emissions. In light of these stark differences, the calls for reparations from leaders across the African continent to Western countries to finance adaptation programmes remain pitiful. Yet, the call on African countries to retract new oil and gas exploration licences is increasing as the results will land in the hands of multinational fossil fuel companies who are set to make billions from oil and gas extraction while the going is good.[11]

Colonial history and present-day corporate imperialism have shown that the wealth generated from resource extraction in many parts of the African continent has not benefited the people in those regions who are directly associated with resource extraction operations. But if you factor in the role and immense profit of Western corporations engaged in exploiting fossil fuels and minerals to chastise the DCR for exploiting their resources, this smacks of (post-colonial) hypocrisy. As cited in the Introduction, for every US$1 the countries in the so-called Global South make, the countries in the Global North make US$30 based on resource extraction and labour to the profits of mineral processing, product manufacturing, and sale. The leader of the Catholic Church, Pope Francis, on a trip to the DRC in January 2023, gave a stinging condemnation of the history of resource extraction from the 19th to the 21st centuries aimed at the various international mining companies. His speech in the capital Kinshasa sent out a clear message to the outside world, declaring: 'Hands off the Democratic Republic of the Congo! Hands off Africa! Stop choking Africa, it is not a mine to be stripped or a terrain to be plundered', telling his audience in this mineral-rich but poverty-stricken country that the diamond industry had 'smeared its diamonds with blood'.[12]

Across the Atlantic in Brazil on the South American continent, during the five-year term of its former President Jair Bolsonaro, 13,000 km² of the country's rainforests were cleared every year through illegal logging, cattle farming, and mining. Destroying vast areas of the Brazilian Amazon (the lungs of the world) not only has global ramifications to the processing of carbon dioxide in the air; it has also decimated the habitats of tens of thousands of indigenous people and their lands. Guaranteed under Brazil's constitution, the Karipuna people's tribal lands in the Brazilian state of Amapá have received no protection from illegal loggers who confiscate their lands and assassinate their leaders. Environmental criminal activity, such as in the Brazilian Amazon where national sovereignty takes precedence over the health of the planet, is both indefensible and abhorrent. Illegal loggers claim they have no other way of making a living with widespread corruption of officials who turn a blind eye, while global environmental bodies bargain with the Brazilian government to set compensation to advert further deforestation.

According to Statista, a global market and consumer data company, as of January 2022 there were approximately 2,250 coal-fired electricity plants in the world. The four leading countries using coal-fired plants are China with 1,110, India 285, the United States 240, and Japan 91. One of the largest exporters of coal, and per capita the largest emitter of greenhouse carbon per population size, is Australia with 24 coal-fired power stations.[13] Nuclear power is again being touted as a clean source of energy production even though the shelf-life of 'spent' fuel rods runs into tens of thousands of years. According to the London-based World Nuclear Association, there are 440 nuclear power plants operating across the world with a further 55 under construction.[14] As of November 2019, the Washington-based Resource Watch mapped approximately 85% of the world's total installed power capacity amounting to 29,000 power plants currently in use. 'Of the plants in the database, 36% of the capacity is coal, 26% natural gas, 19% hydropower, 7% nuclear and 5% oil. Whereas wind, solar and biomass facilities represent 4%, 1%, and 0.6% of the power capacity in the database, respectively'.[15] In a press release from 29 September 2022, the United Nations estimated that 'food loss and waste account for 8 per cent of all greenhouse-gas emissions. By making specific commitments in their updated nationally determined contributions, countries could reduce emissions by 4.5 gigatons of equivalent carbon dioxide per year'.[16] In 2022, 36.6 billion tons of CO_2 was released into the atmosphere making up the bulk of total greenhouse gas emissions to 58 gigatons. The scale that these figures represent is impossible to comprehend as much as the scale of non-renewable to renewable energy production are an illustration of the bandwidth that sustainability policies have to negotiate in order to be successful.

In another recent report in the journal *Science*, David Armstrong McKay lists the six tipping points of irreversible climate degradation on the earth's cryosphere—the total frozen water on the earth's surface. He cites the melting of the Greenland Ice Sheet, West Antarctic Ice Sheet collapse, Barents Sea ice and mountain glaciers melt and the collapse of the Boreal permafrost. Concerning the biosphere, the sum of all the earth's surface and atmospheric ecosystems, climate change devastation includes low-latitude coral reefs, Sahel and the West African Monsoon and Amazon rainforest decimation. The report points to other tipping points ranging from cloud feedbacks due to surface air temperature rises, Tibetan Plateau snow melt, and ocean deoxygenation—'a tipping threshold for weathered phosphorus input to the ocean triggering global ocean anoxia'.[17] Where these tipping points amount to sea level rise, floods, droughts, and devastation of biodiversity, they also tell of the earth's integrated systems that global sustainability has to address. It is clear that in the vast array of the world's ecosystems, climate change is so vastly complex that it is impossible to fully gasp its effects on the planet.

In another study released on 1 August 2022, scientists from the Massachusetts Institute of Technology accused the world and the scientific community at large of not taking seriously the possibility of human extinction from the effects of climate change.

> Prudent risk management requires consideration of bad-to-worst-case scenarios. Yet, for climate change, such potential futures are poorly

understood. Could anthropogenic climate change result in worldwide societal collapse or even eventual human extinction? At present, this is a dangerously underexplored topic.[18]

The authors characterize extreme climate change effects on humans and the planet as a 'climate endgame' leading to mass extinction of human, animal, and plant life if global CO_2 reduction targets are not enforced to avert further global warming. 'It is time for the scientific community to grapple with the challenge of better understanding catastrophic climate change', the authors convey what is now obvious especially to today's angered youth who will inherit the earth. Warnings citing the effects of climate change on the future of human existence on the earth have become so common as to seem fictional rather than real. Raising the alarm to human extinction has often been the stuff of science fiction in portraying the world suffering any number of cataclysmic events. Attaching human extinction to animal and plant extinction is a matter of life and death, yet even this reality remains discordant with human comprehension. What has risen is humanity's belief that technology, big data analysis, and algorithmic innovation will miraculously repair decimated regions and restore the earth's organic life. Given the seismic shift unfolding on the earth's environments in the 21st century, humanity remains in a fog as to how to deal with climate change.

The fogginess that characterizes humanity's comprehension of superiority over all living things is taken further in its most extraordinary accomplishment of leaving the earth. The International Space Station is coming to the end of its life and is earmarked to be decommissioned by 2030. Russia recently announced it will no longer be part of the international space alliance and will build its own. China is assembling theirs—half of which already orbits the earth—and India expects to have its own space station by 2030. Meanwhile, the United States expects to send astronauts back to the moon by 2025, followed by China in 2030 then Japan, South Korea, Russia, India, and the United Arab Emirates. The amount of money spent on a few people leaving the earth runs into hundreds of billions of dollars. Compare this with the measly sums given to renewable energy, land restoration, reforestation, fresh water decontamination, ocean and air purification and it is clear where the priorities lie, creating an estrangement to the earth and a strangeness to humanity's goals.

As Earth time slips into an increasingly unpredictable environmental future, the idea that globally administered sustainable policies can reduce the effects of climate change appears a distant reality when all of the above is taken into account. Humanity's ability to reverse global warming is seeping further from our grasp and the perilous nature facing the younger generation is becoming more insurmountable. Deploying revolutionary tactics to challenge their authority, indigenous First Nation peoples of the Amazon, Canada, North America, Africa, South America, Oceania, and

Figure 1.1 International Space Station passing over Florida, 1 January 1998.

Source: Image courtesy of NASA.

Australia coupled with global movement collectives such as Fridays for Future, Extinction Rebellion, Animal Rebellion, Just Stop Oil, and many other groups are calling the world to action. Over decades of environmental activism, individual climate advocates such as Al Gore and Isra Hirsi (USA), Autumn Peltier (Canada), Greta Thunberg (Sweden), Helena Gualinga (Brazil), Leah Namugerwa (Uganda), and many more individuals across the world are vocalizing, politicizing, and mobilizing against governments who view the economy as a non-negotiable bottom line over sustainability policies. Celebrity environmentalists also lend their fame to climate change action such as Jane Fonda, Leonardo DiCaprio, Joaquin Phoenix, and David Attenborough. The reigning monarch of the United Kingdom and long-standing environmentalist King Charles III introduced at the 2020 Davos Economic Summit his Sustainable Markets Initiative—an alliance with financial institutions and business across the world to utilize a 'set of principles to 2030 that puts Nature, People and Planet at the heart of global value creation'.[19] King Charles has also created what he calls a 'coalition of the willing' to sign up to his Terra Carta, a manifesto to 'reunite people and the planet'. Referencing the 1215 Magna Carta that proclaimed the rights and liberties of people, King Charles' Terra Carta is an agreement to

> rapidly accelerate the transition towards a sustainable future. The Terra Carta offers the basis of a recovery plan that puts Nature, People and Planet at the heart of global value creation—one that will harness the precious, irreplaceable power of Nature combined with the transformative innovation and resources of the private sector.[20]

Creating an inventory of climate activists is somewhat pointless in view of the hundreds of millions of people around the world demanding climate action, and the millions of people changing their lifestyles to remove non-renewal materials, reduce waste, recycle, ride bikes, take public transport, adopt a vegan diet, and commit to solar, wind, and thermal energy production in their homes. It gives an image of hope for a sustainable world, yet the actions required to achieve this environmentally friendly world swing between positivity and negativity, painting a less colourful picture of the earth's future. The activist work of individuals and collectives to 'out' the gross violations committed by unscrupulous oil, gas, and coal corporations, unprincipled populist politicians, and governments is having an unprecedented effect in resetting people's attention concerning climate change. Where powerful entities on one side constantly engage in moving the goal posts through increased controls over capitalism, corruption, and greed, a far greater set of entities and the will of the people are breaking away from this endgame in pushing for a better, equitable, and livable world.

Over the last 50 years, the global oil and gas industries have generated almost US$3bn a day in profits which have exponentially increased since the Russian war in Ukraine.[21] At the end of January 2023, four of the world's biggest oil companies released their financial figures for the previous year. Riding on the back of the war in Ukraine causing worldwide cost of living increases and shortages in energy supplies, all four oil companies posted record profits. US oil giant Exxon's 2022 profits were the largest in its history amounting to US$55.7bn (up from $23bn in 2021), its local competitor Chevon $US38bn (up from $15.6bn) and the British multinational Shell posting $US40bn, the largest in its 115-year history (up from $20bn in 2021) and BP US$28bn (up from 12.8bn in 2021), likewise the largest in its history. All the CEOs of these companies defended their profits as against their investments in new oil and gas exploration costs. Their open announcement to continue to explore and exploit new fossil fuel fields in the face of deepening danger of global warming exposes the hypocrisy sustainability is up against in the global energy market and the shareholders, banks, and governments, who support them. Oil and gas companies reporting immense profits is nothing new and even before these figures were released, UN Secretary General Antonio Guterres declared in a speech given in September 2022 that it is immoral for oil and gas companies to be making record profits from this energy crisis on the back of the poorest people and communities and at a massive cost to the climate. The combined profits of the largest energy companies in the first quarter of this year are close to 100 billion US dollars. I urge all governments to tax these excessive profits and use the funds to support the most vulnerable people through these difficult times.[22]

To counter people's ethical and civil rights, governments and corporations have become adept in misrepresenting let alone listening to the will of the people and organizations such as the United Nations. On the release of Exxon's record profits for 2022, White House spokesman Abdullah Hasan went on the offensive, declaring: 'The latest earnings reports make clear that oil companies have everything they need, including record profits and thousands of unused but approved permits, to increase production, but they're instead choosing to plough those profits into padding the pockets of executives and shareholders'. In response, Exxon boss Darren Woods, in an interview with broadcaster CNBC, said the White House 'needed to "get its facts straight", noting that the firm had continued to spend money on oil and gas projects despite pressure from investors and others to shift investments to renewable energy'.[23] Big fossil fuel corporations have become proficient at shifting the focus away from immense profits and skillfully dodging climate truths while cultivating public ignorance.

Given the serial lies and misquoting by fossil fuel companies, it is beyond comprehension that these companies survive calls for taxing their profits appropriately and proportionally to finance climate change policies. The future of the world has fallen to the responsibility of the powerful few in government and multinationals who decide on the lives and the health of billions of people. The cost of such corruptible power and capitalism must be seen as a failure of humankind. To be against sustainability is not to contradict what it can do but is a response to its abuse and how it is dispensed. It is clear that sustainability can be highjacked, derailed, and diluted by governments, capital markets, and corporations. When the need arises, governments and corporations badge their sustainability credentials to demonstrate that they are committed to reducing carbon emissions to save the planet. Over the 27 years of the COP conferences, procuring sustainability has been the fundamental policy in tackling climate change and global agreements between countries. Over the same period, the world has experienced increased catastrophic weather events, widening the gap between Earth restoration and ongoing devastation. Who is winning the battle concerning climate change?

Besides the fossil fuel industry's effects on global warming and climate change is the construction industry's impact on the earth. From the mining of raw materials to processing, transport, and building, the construction industry is responsible for 23% of global air pollution and 40% of global material waste. It may come as no surprise that concrete is the second most used material in the world and sand, the main material for concrete, second most capital commodity after water. The massive use of concrete and other hard surfaces such as asphalt are not only environmentally damaging due to the sheer number of resources required for their production but also the heat these materials radiate in cities increases the use of artificial cooling and heating. We buy electronic devices and kitchen appliances and have them shipped across the oceans to arrive at our doorstep. In the decades of internet shopping, it is likewise no surprise to buy a hairclip or USB stick for $2 and have it shipped from China to the UK, America, or Europe. We know it is wrong in terms of the CO_2 emissions the hairclip and USB stick create in their passage from factory to door, but we do it anyway. While humankind is living at its peak of consumption and use of non-renewable resources and where talk, policies and practice of sustainability ride high on the agendas of governments, corporations, industry, and the global population, they then waver when significant life changes are called for to reduce human impact on Earth.

There are of course major facets of sustainability that yield positive outcomes for humanity. Global production of food has radically changed through the introduction of genetically modified seeds. Farmers are being encouraged across the world to grow new drought-resistant crops, deploy devices to manage better planting and harvesting methods and better managing of water systems, but with these technologies/methods there comes a cost. The former American agriculture company Monsanto, now part of the German chemical conglomerate Bayer, is one of the largest agrochemical and biotechnology companies in the world. Farmers who use their genetically enhanced seeds are contracted over their lifetime leading to monoculture practices and severe reduction of native and disease-resistant varieties. To sustain genetically modified seeds' high harvest yields requires the use of fertilizers that during heavy rains run-off polluting river systems. Other technological advances include solar, wind turbine, hydrogen, and thermically designed buildings in an effort to 'green' energy production and living. Other new 'cleaner' energy production techniques are being developed such as nuclear fusion which produces limited radioactive waste of far less potency than nuclear fission. Nuclear fusion also provides a far greater amount of energy than nuclear fission where the process of splitting atoms results in long active life cycle of radioactive waste. At present nuclear fusion has its limitations for it requires a great deal of energy to combine atoms to produce a limitless source of energy. Though rising rapidly, renewable energy and negative carbon savings amount to minute percentages when compared with ongoing fossil fuel energy production. This is especially prevalent in many developing countries instigating clean energy production such as solar, wind, and investment in the overall energy sector are faced with paying higher interest rates on borrowing within the international financial market than developed countries. The amount of sustainable energy production required to reduce damage to the natural environment needs to be at a scale that will not be reached until the end of the century. What humanity was able to do in one century—the 20th century—in dramatically impacting the earth, the hope now rests on the 21st century to reverse it. With the population set to increase by 2 billion people by 2050 to 10 billion, economic growth and consumption will likewise need to increase, which will in turn place more pressure on the earth's resources and the present rate of species extinction of animal and plant life. The factors and realities of the 21st century with regard to countering climate change point to a turbulent period in human existence and its ability to reverse what it has created.

The task of achieving sustainability must be embedded in humans before being given over to advancements in technology. Without reconciling the histories of human conflict and suffering, inequality and privilege, global sustainability will remain unachievable. In its present terms of practice, sustainability exists as a silhouette cast onto environmental policies to reduce global carbon emissions and restrict the effects of climate change. The net sustainability of these policies, that is their effectual ability, can be easily disrupted when considering for instance human conflicts as demonstrated by Russia's war in Ukraine. Sustainability has become the go-to silver bullet for achieving the magical number of net zero carbon emissions of the world's leading carbon emitters such as China, the United States, Russia, Europe, Saudi Arabia, Australia, and India. To contain humanity's destructive footprint on the earth, to lift the world out of catastrophic environmental devastation entails addressing issues of continuous economic growth over the health of the planet. The problem widens when sustainability policies are enacted at local, national, and international levels where each level of administration and distribution is open to corruption and dilution. The problem with achieving global sustainability is its remit to solve the slide of the earth's decline by solving humanity's inexhaustible consumption—one wholly endorsed by governments and manufacturers. The gaping hole in the sustainability net cast across the globe comes from its inability to define a common ground between these contradictory forces.

For centuries, the world has been shaped by a minority of powerful industrialized countries and corporations' dominance over the majority of countries and their people. If sustainability is to be successful and the effects of climate change halted, the future of the world must be shaped on ridding the historical racial suppression and resource theft that early colonialists enforced on peoples across the world. This original racial suppression and economic division that would go on to forge global inequality has continued in the post-colonial condition of corporate imperialism by Western multinationals who shape global resources trade to further the economic divide between rich and poor countries. The present ideology and investment being attributed to sustainability is selective and the most vulnerable people affected by climate change are being left out and/or completely ignored. The world continues to be shaped by the wealth comprised of a minority of extremely powerful corporate, political leaders and individuals. Catastrophic weather turbulence, the result of global warming and climate change, is simply a by-product of their maintenance of the global economic status quo.

As outlined in the Introduction, the global pursuit for sustainability cannot be undertaken with shotgun international policy and untested theories that continually fail to reach agreed CO_2 emission reduction targets. The problem with terms and practice concerning sustainability is that they are asked to do everything and for that reason they cannot be radical enough to exercise major industrial and human behavioural changes, to challenge growth economies and the re-election cycles of politics. Citing these issues, the idea is not to undermine sustainability but to reinvent it territorially, culturally, racially as a force of change across the world where territorial boundaries, restrictive migration access, perceived threat, economic protectionism, racial segregation and cultural and religious discrimination are forfeited to the future of human and earth survival—a daunting task given that human history is built on these conflicts. The present situation where the majority of the world's population are far more vulnerable to climate change in comparison to the minority who are the main contributors to global warming due to continued excessive consumption is in fact a rejection of global sustainability. As we are witnessing today increasing numbers of climate refugees seeking to migrate to countries where opportunities for their survival are greater, this shows that if exclusion and inequality continue then mass human mobility and conflict will be forced upon the world and in such a world sustainability cannot flourish.

To speak positively about sustainability in light of what is going on hour-by-hour concerning human impact on the planet overrides what we are doing to save the earth's ecologies. Avoiding the use of plastic, being conscious of recycling comes in the awareness that the sheer volume and energy required to make plastics reusable overwhelms many countries' recycling capacities. Finding new ways to change human habits in order to live sustainable lives requires consistency for their implementation and that consistency does not presently exist in the face of the climate emergency. The regularity of wildfires, land ravished by drought, hundreds of millions of people at risk of starvation in rural areas are a world away from people living in vertical and horizontal planes of urban centres in sealed containers cooled by air conditioners and provided with endless supplies of food stacked on supermarket shelves. The chasm between urban and rural populations' protection and exposure to climate change is extreme to the point where two worlds exist; two realities in non-relational existence to climate change. Keeping global warming to 1.5°C to pre-industrial rise by 2050 is becoming highly unlikely. In 2022, the highest global emissions of CO_2 into the earth's atmosphere were recorded. Achieving carbon neutrality by 2050 is dependent on stopping the emitting and removing of CO_2 from the atmosphere. Various technologies such as 'carbon-capturing' plants are nowhere near the capacity required to reduce CO_2 in the atmosphere, and the costs involved mean that there is a reluctance from both government and industry to finance the thousands of plants required. Of course, one way to finance such technologies is to make the fossil fuel companies turn their huge profits into constructing carbon-capturing plants, removing the carbon from the atmosphere that they put there in the first place.

Climate Adjustment—Combatting Global Inequality

The exodus of early humans from the African continent 100,000 years ago and their walk across the world increased their diversification with each new environment encountered. Through their exposure to different climatic weather patterns, nomadic peoples gathered ecologies, adapting to the environmental conditions along the way. Jumping forward to the 19th and 20th centuries and the combination of industrial–technological–capital proficiency, humanity severed its ecological connections to nature in favour of provisioning its needs. In the 21st century, the accumulative impact of satisfying human needs has resulted in the devastation of whole regions across the world's continents, oceans, north and south Arctic circles, and the displacement of hundreds of millions of people. For humanity to define the basis for creating a global sustainable future is partly dependent on mobilizing the human ecological connections by shifting perspective in accordance with natures. Idealistic but not unmanageable, unearthing human ecology requires a reversing of the historical foundations humanity was built on—resources theft, territorial conflicts, and racial divisions between peoples, cultures, religions, gender, and global inequality that restricts and discriminates against billions of people around the world.

Working out how to achieve a globally sustainable human and earth future involves first dissolving the spatial terms that divide the world's eight billion people into the so-called Global North and Global South. Who fits into the top and bottom of this global compass of division is not solely based on their position on the globe but on the empirical histories and economic dislocations that separate them. The position the 'Global North' occupies on the globe is used to collectively clump the world's rich industrial nations including the United States, Canada, the United Kingdom, the European Union, Russia, Japan, and South Korea. But not all countries of this northern bloc follow the magnetic pull of the compass needle pointing north. Geographically, the Global North gets messy where countries located on and below the equator also make up some of the richest countries in the world, for example, Saudi Arabia, Australia, New Zealand, Singapore, Hong Kong, and developed/developing countries such as China, India, Indonesia, Vietnam, and Malaysia in terms of their average gross domestic product (GDP). The geospatial demarcation messiness is further dislocated in reference to the 'Global South' that namely represents the global poor of the world. Central and South America and Africa fit into this spatial description to convey the poverty of the people living in these regions. While their GDP is low in comparison with the 'Global North', it does not reflect the wealth of their natural resources. For example, the DRC is one of the poorest countries in the world though it is also one of the most resource-rich countries where extraction of its resources is undertaken by multinational Western corporations and aided by government corruption that enriches the 'Global North' by excluding the people of the DRC of a share in the wealth generated. Previous terms used to describe economic differences between nations were given underdeveloped, developing, and developed status. Literally dividing the world's wealth and inequality into north and south can be traced back to the era of colonialism that enriched European countries in the Americas, Africa, South-East Asia, and Oceania. The geospatial divisions concerning rich and poor are a convenient way of clarifying world inequality as much as the continuance of repression, subjugation of labour, resource extraction, and profits created by the Global North at the expense of the Global South.

Implementing global sustainability has to first acknowledge the causes of global division from the colonial period to the present-day disparity concerning CO_2 emissions between north and south. Previously cited but worth repeating the so-called Global North contributes 80% of global CO_2 emissions and the rest of the world the remaining 20%. This disparity is more profound in the case of Africa with its population of over 1.4 billion people in 2022 (17.5% of global population), which is responsible for just 3.7% of global CO_2 emissions. Yet, Africans suffer most from the effects of climate change including severe drought, flooding, heatwaves, diseases due to water contamination, desertification, acidification, and more. The colonial era criminality of the 18th and 19th centuries to the environmental criminality of the 20th and 21st centuries of CO_2 emissions reveal the division of gross inequality concerning climate change between the north and south. Up until now, the world has primarily been unprepared to confront and dissolve these divisions. Huge resources are pooled by rich countries to keep in place their separation from poor countries, and to fuel their resistance to the rights of global human mobility and their deferral of reparations to countries still suffering from the effects of colonialism and cultural destruction. The thought that the atrocities committed by colonialists during their invasion and occupation could be erased over time is having the opposite effect. The ghosts of colonialism have not gone away. Racism, which is a direct result of colonialism, continues to impact on the lives of the most vulnerable people across the world especially when factoring in the effects of climate change.

The United Nations charter is founded on dissolving the world's divisions for the benefit of all humanity. Any number of issues concerning human rights and righting historical wrongs are often blocked by nations who have signed up to the UN treaty, thereby weakening and undermining the institution they established. It is not only about failing to right the wrongs of the past and bring about equality; it is also about removing the embedded system that benefits from inequality. In the course of 400 years, not much was left unpicked by European colonial invaders of First Nations in the Americas, Africa, Asia, Australia, and Oceania. The forced submission of First Nations, the desecration of their lands, the drawing of lines on maps to impose territorial separation, the massacres, slavery, resource plundering, and the cultural and religious persecution have shown, on a monstrous scale, how far racial superiority, inhumanity, and cruelty have guided humanity's preparedness for subjugation and oppression. Though much has changed, 21st century is still dominated by global divisions to what prevailed during hundreds of years of colonial power by one race (white) over black, brown, and tonal skinned coloured peoples of the world. The shameful legacy of colonialism and the human capacity to dominate, oppress, and destroy cultures viewed as 'inferior' still persists in the lack of accountability to address global inequality today. In fact, the traits of colonialism are not just continuing but are deepening in many of the poorer nations of the world. To repeat, global sustainability can only be reached if both past histories and present continuities of the colonialist era are repealed and for countries drained of their resources to be compensated. Only then can the compass be put away and the separation of the Global North from Global South be ended, ridding the world of the divisions of the privileged and the poor, the oppressor and the oppressed.

Climate change has affected poorer nations far more than richer nations. The reason for this is poorer nations are the least developed in securing self-determination, defining national identity, maintaining stable governance, having specifically focused financial institutions, and a lack of infrastructure. The conditions in countries and regions that are environmentally fragile in terms of sustaining human life are exacerbated by weather turbulence created by the CO_2 emissions of their Western counterparts. Sub-Saharan African countries such as Somalia, Kenya, Mali, Chad, and Ethiopia located in semi-arid areas and subject to fragile weather systems are further compounded by the remnants of the colonialist era that saw their dispossession from their lands, decimation of their cultures, and natural practices of land care. In North, Central, and South America, in countries such as El Salvador, Guatemala, Nicaragua, Peru, Brazil, Columbia, Mexico, and more recently Venezuela, the effects of deforestation, flooding, drought, and water contamination are further compounded by violent crime and narcotic street gangs leading to the daily living conditions for the people in those affected regions being tenuous if not lethal. These too are in part remnants of the violence committed during the colonial era, land dispossession, and cultural decimation. Two of the major routes taken by migrants leaving the southern regions of Sub-Saharan Africa and Central and South America to migrate north to reach EU countries and America, respectively, are the combination of the post-colonial condition and present effects of climate change. The recent upsurge in gang violence in the Caribbean Island of Haiti is a prime example of the ongoing effects of post-colonial history of inequality and poverty coupled with climate change.

In *A Brief History of Equality*, Thomas Piketty considers the issue of giving reparations to people and nations who have suffered the immense crimes committed under colonialism where the descendants of those people suffer the effects of climate change today. 'No matter how complex this question is, it cannot be evaded forever: it is time to act, unless we want deep and lasting injustice to continue', Piketty tells us. 'More generally, a colonial heritage of slaveholding forces us to rethink the connection between reparatory justice and universalist justice all over the world'.[24] Piketty gives the example of Haiti, which in 1804 saw the first Black population successfully rise up and gain independence from their French colonialists. To avoid the threat of French invasion to counter the revolution, France placed a huge debt payment for loss of income that the colonialists were deemed to have suffered.

> The 1825 debt, transferred from creditor to another, was officially extinguished and definitely repaid by the beginning of the 1950s. For more than a century, from 1825 to 1950, the price that France tried to make Haiti pay for its freedom had one main consequence: the island's development was overdetermined by the question of the indemnity.[25]

It is no coincidence that Haiti is one of the poorest and most turbulent societies in the world today and the recent and ongoing poverty and violence engulfing the island can be seen as consequence of colonial devised debt enslavement as much as the political assassination of its president. The 1825 debt passed on Haiti amounted to its financial enslavement in making the payments that only ended in the 1950s directly links to the country's lack of infrastructure and ability to combat the devastating effects of increasing hurricanes on the island such as Hurricane Matthew in 2016. The hurricane caused unprecedented destruction, physically wrecking the country and leaving destitute much of its population. Since President Jovenel Moïse's assassination in 2021, gangs now control most districts in the capital Port-au-Prince where kidnappings and murder are everyday occurrences. The civil breakdown of Haitian society in the 21st century is the result of its post-colonial condition of inequality and debt merged with climate change. If nothing more, Haiti's civil collapse is a warning to the rest of the world in which the effects of climate change and colonialism have led to an ungovernable society.

The effects of climate change such as hurricanes, drought, flooding, and societal breakdown leading to gang violence, kidnappings, and gender abuse—the rape of women and girls—are a collision of past and present injustices between humans and the natural world and between humans. The multidimensional issues to which global sustainability is exposed—human conflicts of war and terror, gross economic inequality, selective freedoms of mobility—are fundamental issues that need addressing just as much as those of curtailing catastrophic weather events by reducing CO_2 emissions. Where a quarter of the world's population is exposed daily to food and water insecurity primarily due to climate change, this amounts to an all-out assault on the status quo of rich nations that have kept the disadvantage in check. It is possible to assume that prior to colonialism, the idea of disadvantaged peoples did not exist in the same (institutional) way that it does today. The colonialist construction of making people disadvantaged through the colour of their skin and sense of inferiority is not lost today where the richest 10% of the world's population (800 million) own 76% of global wealth, while 50% of the world's population (4 billion) own 2% of global wealth and the remaining 22% of the world's population (3.2 billion) own the rest.[26] These statistics illustrate the reality that humanity must face in trying to reach a shared sustainability. It is estimated that 3.6 billion people globally live in regions directly affected by climate change and this disparity is an indictment of the inhumanity of humanity.

The task of achieving global sustainability is aimed at reducing human impact across the earth—clean energy, sustainable food production, net zero carbon emissions, reduced consumption, and environmental restoration as a strategy for securing a livable planet for future generations. Farmers across the world understand that if the environment is under threat, their capacity to produce food is threatened as is their survival; for them to succeed, the environment must be protected. As much as this message simplifies sustainability, it is routinely objectified and highjacked by sectors of government and industry who advocate that plundering the earth's resources secures the world's economic growth. Given the sway of such attitudes and acknowledging that people in rich Western nations are unwilling to jeopardize their wealth and comfort, maintaining economic growth in favour of environmental decline weakens the very idea of sustainability let alone its implementation.

Humans have a long history of failing to hold their concentration to successfully solve challenging issues. Whether it be conflicts, natural disasters, resource depletion, CO_2 emissions, our focus shifts to a new set of issues in a perpetual cycle of confrontation. Humans do not necessarily confront issues with the aim of solving them; instead they rely on coming to agreements. In confronting the earth's environmental decline, agreements are made not with the earth but between humans. This could be construed as the same thing, but the position is humans first and the earth second. Human communication with the earth is one Herbert Girardet refers to as the 'process of *entropication*—of combining resources into products and producing wastes faster than they can be converted back into useful resources'.[27] Humanity's willingness to reduce the scale of its impact and consumption requires an immense psychological shift in human relations with the earth's resources. Human willingness to scale down its effect on the earth, animals, plants, and all living things speaks to the immense scale of the task that sustainability must confront.

This devolution of humanity from nature has rendered its present ecological splinter. The relations that once bound humanity to its surroundings and each other as relative, dependent, and equal were shattered in less than 200 years during the 19th and 20th centuries. Humans have learnt that their capacity for destruction is equal to their capacity for creation. 'The power of reflection also constitutes its strength', French theorist Jean-François Lyotard remarked. At present, the power of humanity over the earth reveals a double exposure of environmental and human disappearance. It is clear humans have overreached their scale a billion-fold in relation to the scale and time of the earth. The ratio of an ant able to carry 50 times its own body weight is extraordinary given its physical size. The ratio of human load to impact on the world's ecology is astonishing by its physical destruction. One way to measure the ratio of human impact on the earth and its ability to achieve global sustainability can be seen in the fact that by 2050 the world's garbage output will exceed 3.4 billion tons a year. Presently 90% of human waste is buried in landfills or dumped in nature and the remaining 10% recycled. This is a load the environment cannot bear but it is the hole in which humanity has found itself and out of which it must dig itself.

Sustainability's present model is to reform the sectors that are destroying the earth's biosphere: fossil fuel energy production, deforestation, mining, land degradation, water contamination, ocean pollution etc. But in reality, what sustainability really requires is a revolutionary rethinking of human life on the earth. For sustainability to work as the main driver for securing the future of humanity's existence, it requires all of humanity to become an ecological partner with animal and plant life on Earth (a point further explored in Chapter 4). Terrestrial and mobile, the future of humanity is a wearer of the earth's ecology. Sustainability's starting point is to address climate change, that is, to address everything bad done to the environment by implementing initiatives for doing everything good. Laden as the cure, the load sustainability bears has become its burden. Humans suffer from the same complex. Laden to do everything good to secure the earth's future, humans find ways to circumvent their doing good for that begets *their* burden. Human capacity to carry its load in environmental restoration has only so much tolerance before exhaustion sets in, respite takes hold, and commitments fade.

Action—Climate Resilience and Decarbonisation

Global warming is not only impacting the planet's natural environments; it is also psychologically impacting people's capacity to manage their response. Climate anxiety, trauma, and fatigue are becoming a common part of everyday life for billions of people around the world affected by climate change. Across the globe, young people are battling with the prospect of inheriting an unstable world not of their making. Confronted by the excesses of previous generations' vast consumption of resources without much respect or knowledge of the destruction to the planet being caused and not knowing what the future Earth will be, young people are living under a state of constant anxiety. Intergenerational conflicts are surfacing as the youth of today are angry at the environmental irresponsibility of the older generations who they see as maintaining the global system that brought them wealth and security at the cost of poorer nations and their people. Young people across the world are having to take on the role of rescuing the earth's future by revolutionizing the political, corporate, and global financial systems, demanding renewable energy production, and reducing levels of consumption as an imperative to their future on the earth.

The anxiety young people feel is being reflected across the world. For instance, the words of UN Secretary General Antonio Guterres echoes the anger and anxiety experienced by many today: 'Greenhouse gas emissions keep growing, global temperatures keep rising, and our planet is fast approaching tipping points that will make climate chaos irreversible. We are on a highway to climate hell with our foot still on the accelerator'.[28] Taken from his speech at COP27 conference in Sharm El Sheikh, Egypt, Guterres takes aim at world leaders and fossil fuel corporations who continue to couch mass resource extraction as essential to maintaining the economic health of a nation, stating that such actions are unforgivable and irresponsible particularly to the young and the most vulnerable and impoverished people of the world. The only way out of the crisis Guterres sees is the radical transformation of the global capitalist system, Western governments' wealth protections and multinational fossil fuel corporations' continued exploitation of natural resources, and to right the injustices of the global world order. To achieve climate change resilience is not only to decarbonize industry and reduce consumption; it is to take the youth as key decision-makers in the future direction of the world they are inheriting.

As the human impact on the world's natural environment increases, so too does the sense of anxiety at the individual, societal, and international levels. The effects of climate change are clearly visible, for example, in flooding, drought, deforestation, and desertification. With predicted temperature rises from pre-industrial levels from 1.1°C to projections of 1.5°C, 1.8°C, 2.4°C, and 2.8°C by the end of the century, it is hard to comprehend what this means for the planet other than some vague sense of an approaching cataclysmic future concerning humanity's fate. A psychological disorder is growing across the world concerning how climate change is experienced between the protected and the vulnerable. People living in wealthy countries might feel vulnerable to the future direction of the world, but this radically differs from those people living in regions where their livelihoods are irrevocably impacted and threatened by climate change. Such differences of experience in light of the realities of climate change may explain the sluggish approach to correct this imbalance between those who are (relatively) protected from and those most affected by the effects of climate change. The experiences between the climate change resilient and victim populations will no doubt shift over time but the divisions will remain. It might be the case that climate change

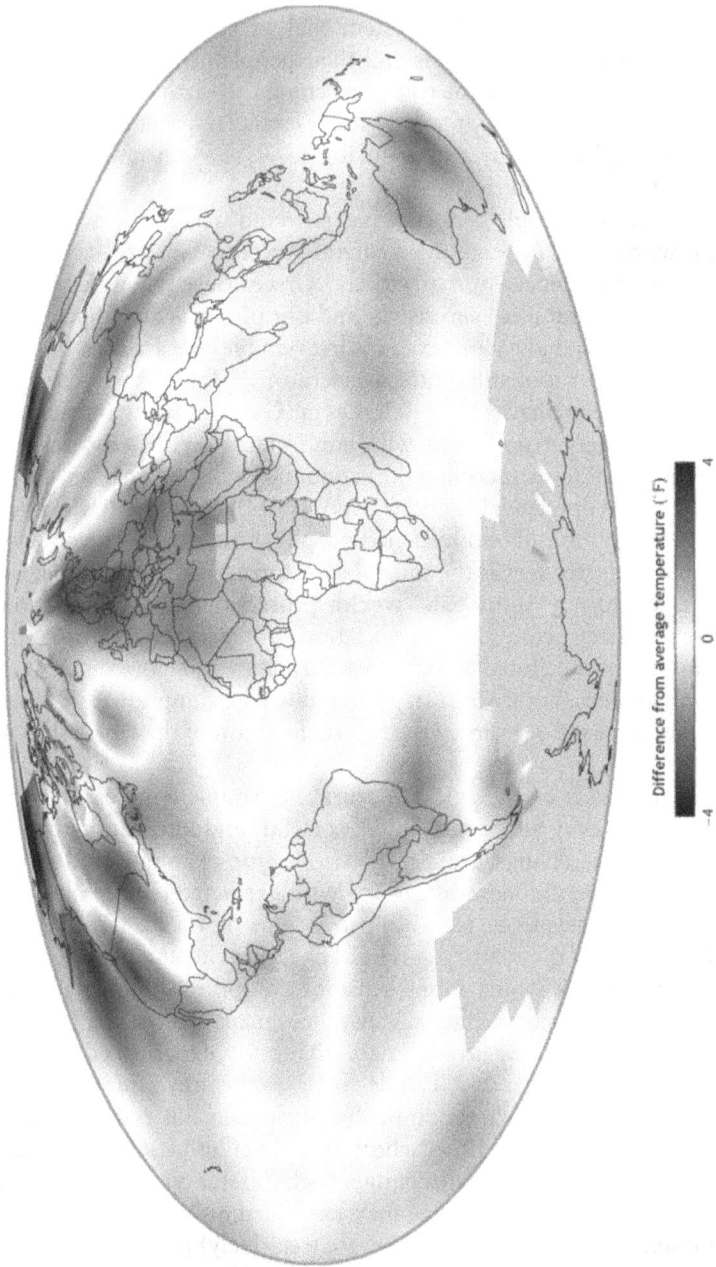

Figure 1.2 Average temperature in 2014 compared with the 1981–2010 average.

Source: Image courtesy of The Decolonial Atlas.

producers responsible for global CO_2 emissions will eventually become its victims and the present victims will perhaps become better equipped to be climate survivors due to their exposure—an outcome that is neither negative nor positive but probable.

Differences in wealthy and impoverished people's resilience to climate change is the focus of the UN-backed Race to Zero. This international organization aims 'to rally leadership and support from businesses, cities, regions, and investors for a healthy, resilient, zero carbon recovery that prevents future threats, creates decent jobs, and unlocks inclusive, sustainable growth'. The organization deploys a multivariable strategy 'of leading net zero initiatives, currently representing 52 regions, 1,122 cities, 7,552 companies, 1,114 educational institutions, 555 financial institutions, over 3,000 hospitals from 63 healthcare institutions and 24 "other" institutions'.[29] In the face of this and many other global organizations' commitment to building climate resilience for the most vulnerable thereby reducing the differential gap, the global financial sector continues to operate with an eye on profit margins and as such is reluctant to change its operations and practices. In an interview at COP27 conference, the UN Race to Zero ambassador Sarah El Battouty complained that the global construction industry, a major contributor to climate change, is 'severely lagging'. Citing two recent UN reports—UNFCCC released on 8 November 2022 and the 2022 Global Status Report for Buildings and Construction: Towards a Zero-emission, Efficient and Resilient Buildings and Construction Sector—El Battouty criticized the construction industry's slow uptake to achieve substantial carbon reduction, especially given that it is responsible for more than 23% of global carbon emissions. 'The entire sector and all its players, including the architects, are lagging', she claims. 'We're not accountable. Nobody says: this isn't an energy-efficient building. Nobody says: why has it been designed poorly?'.[30] Differences in enacting operational changes, whether it be the construction industry or the global financial markets, are determined by the constant drive for growth and profit; as such, the effects of climate change are experienced through the differentiation in the affordability between resilience and victim.

One of the earliest international agreements to combat global warming and reduce greenhouse gas emissions produced the concept of 'cap and trade' in the implementation of carbon trading. The 1992 Kyoto Protocol consisted of an international credit system in carbon emissions trade as a positive and profitable global solution to tackling climate change. An example of trading in carbon emissions can be highlighted whereby one ton of CO_2 is offset by the planting of trees and through the natural process of photosynthesis, carbon is absorbed from the atmosphere, which translates into one carbon credit. Companies with excess carbon credits can trade these with companies who are in a deficit as a means of reducing their carbon footprint. The carbon trading system mirrored the global financial trading system in the trade of commodities on the stock market. While carbon trading looked good on paper, in reality it fuelled a programme of false belief to serve fossil fuel companies where, under this system, continuing their operations could in fact be claimed to benefit the environment! Inside this carbon trading deception, the differential experience between climate resilient and climate victim populations, however, remained the same. The most vulnerable people affected by climate change were excluded from this financial trade in carbon for profit of the planet. For billions of people, carbon trading moved in the ether as container ships moved goods across oceans—it was invisible. The carbon trading scheme was a form of sustainability tailored for the industrial elite who used carbon trading to offset their greenhouse gas emissions while imposing the devastating effects of climate change on the impoverished. Investments in solar, wind power, and tree-planting schemes had little real positive effect on the planet—in essence, carbon trading was sanctioned environmental corruption.

Another United Nations climate change initiative was the establishment in 2013 of the REDD+ reforestation framework, which aimed to restore degraded forests through commercial investments. As was the case in the trading of carbon emissions, companies could invest in forest regeneration projects and/or secure the protection of natural forests from being cut down by funding new avenues for sustainable programmes in agriculture and allied industries. What became of this practice was that investments were often made in existing forest reserves rather than the more costly new reforest-ation programmes, and when they were made, they were not maintained and so this led to sapling failures. As a result, the principal gap ratio between deforestation and reforestation continued to increase.[31] Another carbon trading initiative is the International Emissions Trading Association (IETA) set up in 1999 as a non-profit organization. In its mission statement, it declares that its aim is to: 'Empower business to engage in climate action, advancing the objectives of the United Nations Framework Convention on Climate Change and the Paris Agreement as informed by IPCC science'; and 'Establish effective market-based trading systems for greenhouse gas (GHG) emissions and removals that are environmentally robust, fair, open, efficient, accountable and consistent across national boundaries'.[32] The message of the IETA, namely to tie commercial enterprise to the reduction of greenhouse gas emissions, is another example of placing a value system that is 'robust, fair, open' between carbon reduction and carbon emitters. The traction of IETA trade was the media exposure industries could gain through their association with the scheme. Again, the differential between greenhouse gas producers and climate change victims remained largely unchanged.

It is not just business when it comes to CO_2 emissions 'cap and trade' and climate change. Returning to the 2022 COP27 conference, the gender imbalance of delegates exposed the differentiation in the representation of women and men attending where women involved in the national negoti-ating teams comprised only 34% of the total delegate attendance.[33] While this is an unacceptable statistic, it will no doubt spur a more equal represen-tation by the time of COP28 in 2023. But the real disturbing reality to this imbalance of women's representation in global environmental negotiations is the daily crisis women face due to climate change. A 2022 UN report by UN Women states:

> The climate crisis is not 'gender neutral'. Women and girls experience the greatest impacts of climate change, which amplifies existing gender inequalities and poses unique threats to their livelihoods, health, and safety. Across the world, women depend more on, yet have less access to, natural resources. In many regions, women bear a disproportionate responsibility for securing food, water, and fuel.[34]

The report further identifies the effects on women and girls becoming twice, three times, and four times victims of climate change when compared with men.

> Climate change is a 'threat multiplier', meaning it escalates social, political and economic tensions in fragile and conflict-affected settings. As climate change drives conflict across the world, women and girls face increased vulnerabilities to all forms of gender-based violence, including conflict-related sexual violence, human trafficking, child marriage, and other forms of violence.

The UN Women report makes clear the discriminatory and inequitable nature of climate change negatively impacting on women and girls, which mirrors the discriminatory and inequitable nature of global sustainability policies. Reforming patriarchy, government, society, civil law, cultural attitudes, religious institutions to acknowledge the specific effects of climate change on women and girls is central to reaching sustainability in the 21st century.[35]

People across the world are more aware of their environmental impact and are taking steps to change their lifestyles and behaviours to avert further ecological devastation. While the scale of what is required can feel overwhelming, many people are reducing their carbon footprint through actions such as insulating their homes, installing solar panels, eating less red meat, consuming less, and wasting less. Changes are also afoot in how people are taking public transport and cycling to work rather than driving, conducting business meetings online rather than flying, and buying food and products locally rather than items shipped from faraway places. Across all age groups there is the belief that being more environmentally responsible will positively contribute to the restoration of local, regional, and global ecologies. People are self-organizing, forming collectives to redirect resources, and construct new models of living and working. New off-grid urban and architectural projects utilizing less carbon-based building materials and renewable energy are springing up across the world. Even though fossil fuel consumption is projected to increase over the next three to five years before renewable energy infrastructure projects take effect in phasing-out their use, the motion underfoot is being driven by individuals and communities. One example where people are making a difference can be seen in Berlin's Circular House. Currently under construction, this multi-storey community workspace is being built almost entirely from recycled materials where many of the materials and interior furnishings are sourced from demolished buildings. Overall, the construction industry consumes approximately 50% of all raw materials extracted from the earth and makes up 30% of total global waste. Not including the vast energy requirements for the processing and smelting of ores into metals, composites, and vast amounts of sand for concrete, initiatives like Berlin's Circular House and many other low-carbon building projects underway are seeking to address the industry's excessive carbon emissions. Innovations in building design utilizing parametric modelling, on-site material adaption to changing tolerances, improved passive thermal heating and cooling, green walls, and the use of plantation timber to construct high-rise wooden buildings are making significant inroads. But as the UN Race to Zero ambassador Sarah El Battouty has pointed out, a large proportion of the commercial building industry is still 'severely lagging' in both commitment and scale of implementation.

The rise of green energy production in the form of wind turbines, solar, and to a far lesser extent hydrogen power and electric cars is seen as fundamental in shifting from fossil fuels, but as with anything humans do, it comes with an environmental cost. Producing green energy requires vast amounts of rare-earth minerals to supply battery production for electric cars, wind turbines, and solar panels, which also results in substantial environmental impact. The world's biggest producers of photovoltaic panels, namely China, the European Union, the United States, Vietnam, and Japan, are engaged in a race to control the rare-earth mineral market. Rare-earth minerals such as cerium and neodymium that are essential for electric car batteries, wind turbines, and solar panels, which require vast amounts of water for their processing have decremental effects on the surrounding environment. The toxic discharge of waste water from their processing contains various acids, mercury, and other environmentally damaging substances, which pollute ground water supplies and increase soil acidification.[36] Driving an electric car is a visual sign that a driver is trying to be environmentally responsible, but this overlooks the environmental damage of mining and processing rare-earth minerals in another distant part of the world. Countries like the DRC and Bolivia, which have large quantities of rare-earth minerals, are paying the price of the damage caused to their environment.[37]

After more than a century of fossil fuel dependence comes a new dependence on renewable energy to do the same. The path to global sustainability through energy transition referred to as the 'green revolution' does not take into account increases in traditional mineral mining such as iron ore, copper, bauxite, and gypsum that leave a huge carbon footprint. As the global economy is dependent on capitalism maintaining continual growth, its goal is to increase consumption by making available products in poorer nations without necessarily focusing on decreasing or addressing the foundations of global poverty. For sustainability to be effectual in reversing the effects of climate change, the global economy has to apply 'degrowth' policies that are not consistent with the logic of capitalism. Mining companies are racing to discover new rare-earth mineral deposits where the market is expected to rise 10-fold over the next 20 years. After two centuries of environmental devastation, fossil fuel companies have found a new avenue to continue their domination and expand their portfolios and profits.

Alongside the construction, manufacturing, and renewable industries is the textile industry, contributing to approximately 10% of global CO_2 emissions. The textile industry is third to the fossil fuel and construction industry in terms of emissions output and greater than airline and shipping industries combined. The turnover of fashion, calculated in the short-term life of clothing items, is fed by fashion labels, particularly online brands able to produce a new collection every two to four weeks accounting for between 12 and 24 collections per year.[38] Fifty-six million tons of clothes are sold per year generating $US3 trillion. It is estimated that by 2030 the textile industry's global emissions will increase by 60%.[39] The textile industry uses vast amounts of water for processing and dying. For example, to produce a pair of jeans, from the growing of the cotton to dyeing, requires 3,700 litres of water. While the fashion industry engages in media campaigns to promote that it is embracing sustainable practices, companies often use low-paid workers in poor working conditions. In countries such as Bangladesh, Vietnam, Thailand, China, and India, a great deal of the clothes they make will be worn only a few times or even not at all with most ending in massive dumping grounds. In the Atacama Desert in Chile, clothes transported from around the world have formed a 60,000-ton textile hill. The scale of this and thousands of other clothes dumps around the world is made worse by the use of poor-quality fabrics and cheap manufacturing, meaning that only 1% of clothing is deemed suitable for resale. In Europe alone, four million tons of clothing is dumped each year and only a small portion of this discarded clothing is collected, sorted, and bundled into huge bails by commercial clothing companies and shipped to various developing countries for resale. However, given this small percentage the trade in second-hand clothes accounts for approximately $US64bn a year worldwide. The Kenyan port of Mombasa is a major destination and distribution centre for the second-hand clothing trade supplying many East and Central African clothing retailers. Global marketing campaigns have sought to bring attention to the impact of disposable clothing on the environment and people are becoming more conscious in the purchasing of new clothing. But given that the industry's emissions are set to rise by 60% by 2030, it would appear that the textile industry is far from reaching any sustainable targets.

The extent of the fashion industry's 'manufacturing of desire' for ever-increasing consumption and waste is almost matched by that of the bottled water industry. Throughout the world one million plastic bottles are sold every minute and most of these end up in landfill, lakes, rivers, and the oceans. To counter this plastic waste, scores of new recyclable materials are being developed to replace plastic bottles. Organic compounds such as mycelian (mushroom spores) and biodegradable polylactic acid (PLA), corn starch, and other organic materials for polystyrene food packaging show promise but given the slow roll-out and ability for mass production, their effectiveness is miniscule. For the foreseeable future, plastic water bottles and food containers will continue to be thrown away in their hundreds of millions a year, filling waste dumps, clogging waterways, and seas. Reforestation is another global programme aimed at tackling climate change. According to the United Nations Global Forest Resources Assessment 2020 report on deforestation, 10 million hectares of the world's forests are lost every year and deforestation accounts for 20% of global carbon emissions. According to the report, 'global total forest area stands at some 4.06 billion hectares but continues to decrease'. The Food and Agriculture Organization (FOA)

> estimates that deforestation has robbed the world of roughly 420 million hectares since 1990 mainly in Africa and South America. The top countries for average annual net losses of forest area over the last 10 years include Brazil, Democratic Republic of the Congo, Indonesia, Angola, Tanzania, Paraguay, Myanmar, Cambodia, Bolivia and Mozambique.[40]

When two of the world's largest rainforests—the Amazon Forest and the Congo Basin—have undergone marked increases in deforestation, new measures for reforestation to restore degraded farmland are being introduced. As previously mentioned, tree planting is one way people can get actively involved in restoring degraded land and reduce CO_2 in the atmosphere. The survival and failure rate of tree-planting projects varies from country to country and region to region where more than 50% of saplings may die. This was the experience of Tony Rinaudo of the Farmers Management Natural Regeneration (FMNR), part of the UN World Vision programme. In many of the reforestation projects he undertook in the West African state of Niger, most if not all his tree sapling plantings died due to lack of water, animals, and disease. A change in tactics came when he noticed that in severely degraded land areas, the original tree stumps left in the ground after clearing were sprouting clumps of shoots. By pruning these shoots and leaving the strongest to grow, trees grew back without a single sapling being planted. Tree regrowth also brought increased crop yields for they assisted in retaining moisture in the ground, reducing soil erosion, and providing natural soil fertilization. Since 1983, 200 million trees in Niger have regrown and tens of thousands of hectares regenerated through this method. In the last 10 years, the FMNR programme has spread to over 13 countries in Africa and Asia remediating land reforestation covering 1 million hectares and improving the food security and wellbeing of more than 6 million people.[41]

Back in Tony Rinaudo's homeland of Australia, reforestation is also on the agenda in severely degraded regions. In Western Australia, 10,000 hectares of degraded semi-arid farmland is undergoing restoration through the planting of 20 million trees, while on the other end of the continent, in the Gulf of Carpentaria 40 million mangrove trees perished along the coastline as a result of the 2015 El Niño effect. Extremely dry weather and a drop in sea levels left the mangrove forest exposed due to the lack of water. Attempts to restore the mangrove forest are currently underway but the remoteness of the area and ongoing dry weather due to climate change are hampering the progress. On the African continent, the Ethiopian government has created a target of planting four billion trees in a single year. Set in 2019, the target was supported by reports claiming the planting 350 million trees in a single day. If such numbers are true, what is more important is the success rate of tree saplings taking root and flourishing and, taking the two-year Tigray conflict 2020–2022 in the north of the country into account, verifying the success of the programme is difficult.[42] The Australian and Ethiopian examples are a small part of a larger worldwide reforestation programme in an attempt to counter the far larger rate of global deforestation undertaken by humans. It illustrates human resilience and proactive schemes in decarbonizing the planet. But the huge task in reducing deforestation and increasing reforestation to restore photosynthesis processing of CO_2 from the atmosphere runs up against the barriers of human greed and corruption of illegal loggers, cattle farmers, and officials.

Another way humans have resorted to adapting to climate change is to tinker with rain making. Cloud seeding—the process of seeding clouds with compounds to produce rain—has been around for decades. Known as flaring, silver iodine is released from planes into clouds making them heavier so as to precipitate rain. Approximately 60 countries are involved in cloud seeding on a regular basis and this highly controversial practice brings its own conflicts to adjoining countries. The controversy surrounding cloud seeding is at the centre of nation-state atmospheric controls. As clouds travel over regions, countries, continents, and oceans, it is not possible for any one country to claim ownership of them but it is possible to appropriate them. What cloud seeding essentially allows is the ability to steal clouds from other countries and (re)programme them to rain in designated areas. For example, Israel seeds clouds that would otherwise move by natural cause of airflow to Jordan. The same goes for the United States and China where cloud seeding reduces the natural flow of clouds to Mexico and Vietnam, respectively. China spends the equivalent of $US200 million seeding 50 billion m³ of cloud cover a year with the capacity to have some impact over two-thirds of the country's overall rainfall. At present there is no accountability over cloud seeding between adjoining countries leading to environmental conflict over the control of rain. Like the wild west of oil exploration at the beginning of the 20th century, cloud seeding is the new wild west in the skies, taking on a geo-thermal criminal dimension where one country accuses another of 'cloud stealing' coining the term 'rain thieves'. Cloud seeding and cloud stealing constitute one of many appropriations of environmental controls being fought between countries concerning oceans, ice shelves, the North Pole, and Antarctica, posing challenges to international environmental laws and justice on environmental manipulation of the atmosphere and weather.

At a global policy level various resolutions made at the 27 COP conferences since 1995 have often failed to live up to the agreements signed by the participating countries. At the 2009 COP15 conference in Copenhagen, rich Western nations agreed to support those nations most vulnerable to the effects of climate change to the tune of 100 billion dollars per year to be reached over 10 years ending in 2020. By the end of the period, a total of $US80bn had been reached over the whole 10-year period. The 100 billion dollars a year is now set to be reached in 2023. The 2022 mass flooding in Pakistan resulted in one-third of the country being under water, causing damage estimated at 40 billion dollars. At the time of writing, Pakistan had received $US816 million from the United Nations up from the previous $US160 million first pledged.[43] At the 2022 COP27 conference in Sharm El Sheikh, an agreement was reached after a great deal of sustained pressure from vulnerable nations on the world's richest nations to fund countries most at risk from the effects of climate change. Known as the Loss and Damage fund, the details of the deal appear to make good on the $US100bn a year climate funding target agreed back at the 2009 COP15 conference.[44] The Loss and Damage fund is geared to finance climate adaptation for the most at-risk nations, for example, Madagascar, Afghanistan, Bangladesh, Pakistan, Malawi, Chad, Kenya, and Haiti falling victim to drought, flooding, extreme weather events such as devastating hurricanes, but also compensation to countries where their economies are dependent on tourism such as the Maldives and Jamaica where rising sea levels are eroding beaches and natural habitats.[45] While these initiatives speak to the possibility of climate change resilience through adaptation programmes, they are hampered by making good on the agreements concerning their financing.

Currently only 34% of climate financing goes to helping countries adapt to climate change and a large proportion of the climate financing (71%) comes in the form of loans making those nations most vulnerable to climate change, notably underdeveloped countries, fall further into debt from Western loans. At the COP27 conference, Western nations pledged an extra $US230 million to the Adaptation Fund—a worldwide NGO headquartered in Washington, DC.[46] This amount, which is paltry when compared with the devastating impact facing nations of the 'Global South', indicates the realities of the (non-)commitments made. At COP27 conference, there was a great deal of criticism by climate change policymakers at the large presence of fossil fuel lobbyists, and deals to 'phase-out' the use of fossil fuels were reframed to 'phase-down of unabated coal power', which was construed as protecting 'business as usual' for the fossil fuel companies.[47] Separate to the Loss and Damage and the Adaptation Fund is the Global Centre of

Adaptation headquartered in Rotterdam. Founded on 18 September 2018, the centre states that its objective is to mediate policies focused on supporting solutions for environmental adaptation. Its funding sources come from government and private sectors and one of its major programmes, the African Adaptation Acceleration Program, recently received a €110 million boost from the Dutch Government.[48] The centre's focus is directed at financing global adaptation projects in climatically challenged regions affected by climate change. While adaptation projects are contingent on funding from mostly rich Western nations, overall pledges have fallen short due to debts rung up during the COVID-19 pandemic, global financial instability, and the rising cost of living leading to financial commitments being reduced or retracted entirely.[49] Another international network dedicated to tackling climate change is the Green Climate Fund (GCF) headquartered in Songdo, Incheon City in the Republic of Korea. Established in 2013, the fund receives financial contributions from governments, institutions, and the private sector worldwide. Working in more than 100 countries, the aim of the fund is to 'support paradigm shifts in both climate mitigation and climate adaptation efforts. GCF aims for a 50:50 balance between mitigation and adaptation investments over time'. With total financial contributions up to 2022 at $US11.4bn, the GCF has assisted in creating 'increased resilience for 665.9 million people' and achieving '2.4 billion tons of CO_2 equivalent avoided' through the 177 projects it is currently implementing, though it acknowledges by its own estimates that 'developing countries need USD 2–4 trillion annually to avert catastrophic climate change'.[50] This amount is staggering in comparison with all the other amounts previously mentioned. The takeaway from all of these funding organizations mentioned and the hundreds not mentioned as well as the work of climate activists, indigenous environmental knowledges, NGOs, governments, private sector, business, and the general population throughout the world paints a positive future projection of the earth but one held entirely in the clutches of 'environmental charity'—a most vulnerable liability.

I started this chapter with some scepticism concerning the policies and implementation of global sustainability. The repeated use of sustainability has lent itself to being misused, greenwashed, and manipulated as the 'hook' to arrest climate change. The essence of sustainability as policy and implementation swings in a perpetual back and forth pendulum between the divisions of the convenient descriptions of the so-called Global North and Global South, highlighting the vast chasm between the climate protected and climate vulnerable nations. Climate scientists believe that only a shared ecological global approach can stem the devastating consequences of human impact on the earth, something that governments, capital, and fossil fuel industries are well aware of but underhandedly obstruct in order to maintain the global status quo, power, trade, and profit. The vocal climate deniers are deaf to the undeniable scientific evidence of global warming and climate change. It is the young generation and generations to come who will be spending their lives dealing with the effects of turbulent weather events in a radically altered Earth. In the wake of the Russian war in Ukraine, worldwide military spending has increased dramatically—the trillions of dollars from some of the big spenders, including the United States, China, Russia, the United Kingdom, France, Turkey, Israel, Saudi Arabia, Germany, Canada, Australia, and Iran, flies in the face of climate change spending counted in the billions. The victim of this massive imbalance is the attempt to reach global sustainability and protect the quality of human life and all living things on the earth. Obtaining climate justice can only be assured when human justice is guaranteed, and humanity is far away from guaranteeing this. The goal of sustainability is to give justice to global economic equity, equal opportunity, gender equality, climate change victims, freedom of mobility, and self-determination as an equal part to reducing CO_2 emissions, land restoration, environmental adaptation, and technological innovation—it is all part of the same goal of global sustainability.

The effects of climate change are being felt by billions of people throughout the world. Where at any one time 300 million people around the world seek refuge, 2 billion people experience food and water insecurity, and the richest 1% produce more carbon emissions than the whole of Africa, the basics of sustainability in tackling climate change must be based on equality for all people across the world. But if humanity continues to fail to secure basic human rights of food security, health, education, work, and freedom of mobility to all people, sustainability will likewise fail. We presently live in a world where products enjoy greater rights, freedom of movement, and access to the world than many humans. Returning to Carl Sagan's 1985 testimony before the US Congressional hearing on the 'Greenhouse Effect', he warned US lawmakers of the unfolding and devastating effects of global warming 'if nothing is done'. Nearly 40 years later it is impossible to escape from the terms of global warming, climate change, and sustainability. Every news bulletin, newspaper, radio talk show, documentary, film, and everyday people's conversations confirm Sagan's words that 'we are all in this greenhouse together'.

Notes

1 A full transcript and recording of Carl Sagan's testimonial to the US Congress can be viewed on the C-Span site: www.c-span.org/video/?125856-1/greenhouse-effect
2 Directed by American Davis Guggenheim, *An Inconvenient Truth* is a 2006 Oscar-winning documentary film about former United States Vice President Al Gore's global campaign to raise awareness of the effects of global warming on the planet.
3 Sagan created three graphic images to show three epochs of the earth's formation. The graphics were accompanied by a 'welcoming message' to any alien life who might stumble upon the LAGEOS satellite. The short text reads:

> A few hundred million years ago the continents were all together, as in the top drawing. At the time LAGEOS was launched the map of the Earth looks as

in the middle drawing. Eight million years from now, when LAGEOS should return to Earth, we figure the continents will appear as in the bottom drawing. Yours truly.

(See Carl Sagan's book *Murmurs of Earth: The Voyager Interstellar Record*. New York: Ballantine, 1978)

4 For the United Nations press release of 6 June 2022 concerning the effects of Russia's invasion of Ukraine, see www.fao.org/newsroom/detail/un-report-global-hunger-SOFI-2022-FAO/en

5 For a report, see www.dw.com/en/european-commission-declares-nuclear-and-gas-to-be-green/a-60614990

6 This was Fumio Kishida's inaugural address as Japan's PM delivered on 14 October 2021. See www.ft.com/content/ffa6754f-3c12-4729-921d-aa2acc5e96ee

7 See Adrian Parr, *Hijacking Sustainability*. Cambridge, MA: MIT Press, 2009, pp. 1–2.

8 Ibid., p. 30.

9 Ibid., p. 150.

10 See the Wilson Centre report of 24 March 2021 at www.wilsoncenter.org/article/battle-earths-climate-will-be-fought-africa Adding to the Wilson Report is a list of the world's top 10 CO_2 total emissions as of 2022: (1) China, 10,065 million tons of CO_2 released; (2) United States, 5,416 million tons; (3) India, 2,654 million tons; (4) Russia, 1,711 million tons; (5) Japan, 1,162 million tons; (6) Germany, 759 million tons; (7) Iran, 720 million tons; (8) South Korea, 659 million tons; (9) Saudi Arabia, 621 million tons; and (10) Indonesia, 615 million tons of CO_2. Australia does not make it to the list but as a global supplier of fossil fuels and the world's biggest CO_2 emitter per population, it deserves to be included. *Source*: https://climatetrade.com/which-countries-are-the-worlds-biggest-carbon-polluters/

11 Post-colonial independence and security and corporate imperialism are combining to relaunch oil and gas partnerships with oil companies. For example, Tanzania and Uganda are developing a joint project with French oil giant Total to drill for oil near Lake Albert of Uganda and build a heated pipeline covering 1,443 km to the Tanzanian port of Tanga on the Indian Ocean. Citing multiple human rights and environmental risks, the EU Parliament has voted to pass a resolution calling on both countries and Total to halt the oil and gas exploration where more than 400 wells are to be drilled in protected natural areas. Tanzania's Energy Minister January Makamba and the Ugandan Parliament has criticized the EU decision citing EU's attitude as hypocritical and neo-colonialist. Uganda, Tanzania, and Total will invest 10 billion (currency?) into the project which is forecast to be up and running and producing 190,00 barrels per day or 1.4 billion barrels in total till 2050. For more information as well as to object and donate to block the joint venture, see www.inclusivedevelopment.net/cases/east-africa-stop-the-east-african-crude-oil-pipeline/

12 Note: In an article (31 January 2023) for the BBC, Aleem Maqbool made the comment that in Pope Francis' strong message to international corporations profiteering from unabated resource extraction in the war-torn DRC, he did not include the history of the Catholic Church's role over the last two centuries in profiteering from colonial rule and cultural decimation through its missionaries,

writing: 'However, the Pope did not specifically refer to the role played by Catholic colonisers, backed by historic edicts from the Vatican, and the atrocities they committed here'. See www.bbc.com/news/world-africa-64476624

13 To view the complete report by Statista data analysis on worldwide coal-fired electricity plants, see www.statista.com/

14 For a comprehensive listing of global nuclear power plants, see www.world-nuclear.org/

15 For more information on global energy production and resources, see https://resourcewatch.org/

16 See United Nations press release 29 September 2022:
https://press.un.org/en/2020/dsgsm1465.doc.htm#:~:text=Food%20loss

17 For more information on David Armstrong McKay's six tipping points of irreversible climate degradation on the earth's cryosphere published in *Science*, 9 September, Vol. 377, Issue 6611, see www.science.org/doi/10.1126/science.abn7950

18 Further in the report the authors discuss what global warming means to human and earth life and how the global challenges to reducing greenhouse gas emissions are besieged by the actual ability to carry them out.

> Recent findings on equilibrium climate sensitivity (ECS) underline that the magnitude of climate change is uncertain even if we knew future GHG concentrations. According to the IPCC, our best estimate for ECS is a 3°C temperature rise per doubling of CO_2, with a 'likely' range of (66 to 100% likelihood) of 2.5°C to 4°C. While an ECS below 1.5°C was essentially ruled out, there remains an 18% probability that ECS could be greater than 4.5°C. The distribution of ECS is 'heavy tailed,' with a higher probability of very high values of ECS than of very low values.
>
> (www.pnas.org/doi/full/10.1073/pnas.2108146119 Climate Endgame: Exploring catastrophic climate change scenarios, *Proceedings of the National Academy of Sciences (PNAS)* report, 1 August 2022, by Luke Kemp, Chi Xu, Joanna Depledge, Timothy M. Lenton, Massachusetts Institute of Technology, Cambridge, MA)

19 For more information on the Sustainable Markets Initiative, see www.sustainable-markets.org/

20 For more information on King Charles III's Terra Carta, see www.sustainable-markets.org/terra-carta/

21 Responding to the war in Ukraine, the 13 members of OPEC (Organization of the Petroleum Exporting Countries, which includes Russia) announced on 4 August 2022 that they would increase their output by 100,000 barrels a day. This is on the back of the 600,000 barrels increase in July and August. OPEC oil members, who collectively produce 28 million barrels of crude oil a day, have come under pressure from oil-importing countries to increase overall output to bring the price of crude oil down, since prices have soared due to the ongoing war in Ukraine.

22 UN Secretary General Antonio Guterres made this announcement in his speech in the Review Conference of the Parties to the Treaty on the Non-Proliferation of Nuclear Weapons at the United Nations in New York City on 1 August 2022.

23 See BBC report, 31 January 2023: www.bbc.com/news/business-64472806
24 See Thomas Piketty, *A Brief History of Equality*. Cambridge, MA: Harvard University Press, 2022, p. 68.
25 Ibid., p. 73.
26 These statistics were released from the World Economic Forum on 10 December 2021: www.weforum.org/
27 See Herbert Girardet, *Regenerative Cities*. Hamburg: World Future Council and HafenCity University, 2010, p. 6.
28 To hear the full opening address to the COP27 at Sharm El Sheikh by the UN Secretary General Antonio Guterres, see www.youtube.com/watch?v=cIvfekhiY
29 For more information on Race to Zero, see https://climatechampions.unfccc.int/the-race-to-zero/
30 For extracts of Sarah El Battouty's interview, see architectural projects review site Dezeen: www.dezeen.com/2022/11/18/un-climate-ambassador-sarah-el-battouty-interview/. With regards to the two UN reports cited, see UNFCCC report: https://unfccc.int/documents/614385 and the 2022 Global Status Report for Buildings and Construction: Towards a Zero-emission, Efficient and Resilient Buildings and Construction Sector, https://wedocs.unep.org/handle/20.500.11822/41134
31 For more information of REDD+ activities, see https://redd.unfccc.int/
32 For information on the International Emissions Trading Association mission and objectives, see www.ieta.org/About-IETA
33 *Source*: www.bbc.com/news/science-environment-63636435
34 See the full report by UN Women in relation to climate change: www.unwomen.org/en/news-stories/explainer/2022/02/explainer-how-gender-inequality-and-climate-change-are-interconnected
35 Further to the UN Women report on gender equality concerning the effects of climate change, it states:

> When disasters strike, women are less likely to survive and more likely to be injured due to long-standing gender inequalities that have created disparities in information, mobility, decision-making, and access to resources and training. In the aftermath, women and girls are less able to access relief and assistance, further threatening their livelihoods, wellbeing and recovery, and creating a vicious cycle of vulnerability to future disasters.
>
> (www.unwomen.org/en/news-stories/explainer/2022/02/explainer-how-gender-inequality-and-climate-change-are-interconnected)

36 More than 70% of the world's cobalt is extracted from mines near Kolwezi in the DRC. Cobalt is a sought-after mineral essential to battery production for the electric car industry. Transnational mining companies from China, UK, and USA engaged in cobalt mining are making multi-billion-dollar profits, while the Congolese workers employed fall further into poverty due to poor pay and the rising cost of living. Human rights campaigners are calling on companies to increase the pay for impoverished miners, but so far nothing has been done. Cobalt mining in the DRC highlights the modern-day takeover of resources and corporate imperialism sweeping across the African continent that not only connects to its colonial history but is an extension of it. See www.newyorker.com/magazine/2021/05/31/the-dark-side-of-congos-cobalt-rush

37 China currently holds 75% of known rare-earth minerals. In the regions where these minerals are mined and processed, it has led to massive environmental damage, acidification of land and ground water, and toxic air pollution. In Bolivia where roughly 25% of the world's known lithium resources (a core mineral for battery production) are found on the vast Salar de Uyuni salt flats, the economic benefits of its mining are yet to translate to overall wellbeing of this poor nation and where the environmental damage is yet unknown. These are just two examples illustrating the 'sustainability' of the 'green revolution'.

38 The global Spanish brand Zara produces 65,000 products per year, which amounts to 200 new products per day. As one of the first companies to promote fast fashion it has seen its profits rise to three billion a year.

39 For more information on the global emission of the textile industry, see the UN Climate Change report: https://unfccc.int/news/fashion-industry-un-pursue-climate-action-for-sustainable-development

40 At the 2021 COP26, more than 100 of the world's leaders pledged to end deforestation by 2030. The pledge which amounts to 19.2 billion dollars from public and private funds includes the world's most threatened rainforests: Brazil's Amazon and the Congo Basin in the DRC. As deforestation leads to the depletion in processing CO_2 gas through plant absorption, it is crucial to halting further global warming and catastrophic weather turbulence. The countries that have signed the pledge represent 85% of the world's natural forests. All previous agreements negotiated at past COP summits have failed to halt deforestation. For more information on the 2020 United Nations report on global deforestation, see https://news.un.org/en/story/2020/07/1068761

41 For more information on the Farmers Management Natural Regeneration (FMNR) reforestation programme in Niger, see www.worldvision.com.au/global-issues/work-we-do/famine/aussie-spirit-fuels-africas-greatest-transformation

42 Concerning the Ethiopian government's claim of planting 350 million trees in a single day, see the UN press release on 2 August 2019: www.unep.org/news-and-stories/story/ethiopia-plants-over-350-million-trees-day-setting-new-world-record

43 In a recent protest action by the UK-based climate action group Just Stop Oil, calling for the government to end all new oil and gas projects immediately and phase out fossil fuel dependency, activists voluntarily gave themselves up to police after digging a tunnel under a major delivery route to a nearby oil terminal and occupying it over 13 days in August/September 2022. The protesters cited the devastating floods in Pakistan in 2022 as a direct consequence of climate change, where one-third of the country lay under water and tens of millions of people were made homeless and destitute, as one of the reasons for their activism. As a result of the damage caused by the 2022 floods, it is estimated that the country will need US$16bn for rebuilding for the loss of two million homes, with tens of millions of people displaced. Pakistan contributes 0.6% to global emissions, yet ranks seventh in the world as the most vulnerable to climate change.

44 For more information concerning the COP27 'Loss and Damage' deal, see https://climatenetwork.org/2022/11/20/landmark-decision-at-cop27-to-set-up-loss-and-damage-fund/

45 Before the COP27 Conference in Sharm El Sheikh, there was a climate summit meeting of African leaders held in the Netherlands in September 2022 in which

only one Western leader, the Dutch prime minister, attended. There have long been calls from African leaders to Western countries to compensate African countries who are most acutely feeling the effects of climate change but who have the least impact on the earth compared with Western countries according to Congolese President Félix Tshisekedi who noted that The Horn of Africa is suffering its worst drought on record with millions of people on the brink of famine due to the effects of climate change. Senegalese President Macky Sall called out Western leaders' absence and failure to offer funds to African countries to adapt to global warming, describing it as an absolute failure of responsibility considering Western nations are the world's major CO_2 polluters, while Africa is the smallest contributor to climate change and yet suffers its worst consequences. www.businessghana.com/site/news/politics/270227/African-leaders-criticise-West-for-climate-summit-snub

46 For more information of this NGO, see www.adaptation-fund.org/

47 In total, approximately 35,000 people attended the COP27 in Egypt. The size of each country's delegation varies and represents both climate activists, scientists, NGOs, individual governments and leaders, and fossil fuel industry lobbyists. What has increased is the last group—fossil fuel industry lobbyists; at COP27 there were approximately 600—a 25% increase on COP26 in Glasgow the year before prompting one delegate to cynically remark that 'COP27 looks like a trade show for the fossil fuel industry'. At the same time African leaders are asking for climate change compensation and also defending increased oil and gas exploration, boldly pointing out that investment in these industries will help to finance infrastructure such as electricity where 600 million people do not have access to electricity.

48 For more information on the African Adaptation Acceleration Program, see https://gca.org/programs/aaap/

49 This does not mean pledges are not made by some of the world's biggest contributors of global CO_2 emissions to make finances available to nations severely affected by climate change. In 2022, pledges from the USA amounted to $US11bn, the EU €24bn, and the UK £11.6bn. Given these countries are in the top bracket of the world's largest emitters per population of CO_2 it makes sense that they are the largest contributors. But the largest emitter of CO_2, China, and increasingly growing India along with Brazil (calculable by its size of deforestation), who have developing-nation status, are under no obligation to provide financial assistance to other countries for climate change adaptation programmes.

50 *Source*: www.greenclimate.fund

Bibliography

Girardet, Herbert. *Regenerative Cities*. World Future Council Hamburg: HafenCity University, 2010.

Guterres, Antonio. *UN Secretary General Speech COP27, Sharm El Sheikh, Egypt*: www.youtube.com/watch?v=cIvfekhiY

Kemp, Luke, Chi, Xu, and Joanna, Depledge, et al. 'Climate Endgame: Exploring Catastrophic Climate Change Scenarios'. *Proceedings of the National Academy of Sciences (PNAS)* 119, no. 34 (2022).

Parr, Adrian. *Hijacking Sustainability*. Cambridge: MIT Press, 2009.

Piketty, Thomas. *A Brief History of Equality*. Cambridge, MA: Harvard University Press, 2022.

Sagan, Carl. *Murmurs of Earth: The Voyager Interstellar Record*. New York: Ballantine, 1978.

World Economic Forum. 10 December 2021, www.weforum.org

2 Terrestrial Migrations
Nomadic Ecologies

Roaming—Earthly Freedoms

The earth's weather turbulence, we are told, is 'real' and 'permanent', and radical changes are required to prevent it from increasing. Yet this does not reflect the true scale of the crisis. Climate change affects every part of the planet, but its effects on humans and regions across the world's continents are different. The spate of devastating weather events such as drought and flooding in Europe and North America in 2022 wrought destruction across the continents, but a far more significant environmental destruction has been taking place across vulnerable regions in the African, South American, Asian continents and Oceania. The mass exodus of people leaving their homelands has resulted in the displacement of tens of millions of people. This enforced human mobility is prompting regional, national, and international governments, institutions such as the UN and NGOs to deal with the problem, which is exponentially rising year on year. Nowhere is this more visible in South and Central America and Africa where migrants are combating the restrictions placed on their freedom of movement to stem their flow through the militarized national borders of the United States and the European Union struggling with the mass exodus of people leaving their homelands. While climate change refugees are not the sole body making up global human mobility (which includes economic migrants, people seeking refuge from religious, sexual persecution, and gang violence), their collective migration is part and parcel of worldwide inequality and colonial histories that connect to the broader effects of the climate crisis.

In April 2020, a human caravan of approximately 3,000 mostly Hondurans from Central America left their country to journey north with the goal of reaching the United States and crossing the border. The route these people walked was governed by an invisible compass pointing north and the hope of securing a better future. Their unwavering task and risk to their lives was countered against what they had left behind: the daily hardship of poverty, gang violence, gross inequality, and government oppression. Along the way they ran the risk of extortion and kidnapping by gangs, women

DOI: 10.4324/9781003382515-3

and girls being sold into sex slavery, suffering infectious diseases and harassment by policing authorities. They walk on uneasy ground, yet each step they
take fortifies their mission and self-determination. The ground they walk
over moves within them and the journey is written onto their bodies. The
Honduran caravan is a faction of the global exodus of people seeking refuge
from the ravishes of poverty and climate change. On the African continent,
hundreds of thousands of people are also moving north, while millions are
displaced and interred in refugee camps. Embarking on life-threatening
journeys through the deserts of Mali, Niger, Chad, Algeria, Morocco, and
Libya, they move north to cross the Strait of Gibraltar into Spain, cross the
Mediterranean Sea to southern Italy crossing in small rafts with the aim of
reaching Germany, the United Kingdom, and Scandinavian countries. On
the Asian continent, millions of Syrian, Bangladeshi, and Afghan people
have migrated from their countries escaping war, poverty, and oppression to
reach Europe via Turkey to Greece. Many people languish in dire conditions
within detention centres, fall to the mercy of people smugglers, are forced
into slave labour, or drown trying to cross the Mediterranean Sea in unsea-
worthy inflatable rafts capsizing in the rough seas. Despotic leaders, religious
wars, and cultural discrimination mixed with the effects of climate change
are creating this new phase of human mobility. Not seen since the early colonial period of European migration to the lands of the 'new world' and post-
war mass migration, this new evolutionary migration also connects to the
beginnings of humanity's diversity of race and culture that emerged from
Africa 100,000 years ago; first to Asia, then later to Europe 45,000 years
ago and far later 15,000 years ago to the Americas. Of all these migrations it
was early human migration that first exposed humans to the diversity of the
earth—tropical forests, desert, heat, verdant lands, ice flows, cold, prairies,
valleys, mountains, oceans, and islands—in order to forge intrinsic relations
for an embodied human nature.[1]

What links the human caravan of people walking out of Central America
to reach America, and African migrants making similar attempts to cross into
Europe is that they enter a battleground facing expulsion by the authorities
they encounter. A case in point is the Spanish enclave of Melilla located on
Morocco's northern coastline. On 24 June 2022, 2,000 migrants attempted
to cross into Melilla—one of two EU borders on the African continent—
but were violently repelled by the authorities. The second EU border on the
African continent circles the Spanish autonomous city of Ceuta where on
27 April 2021 approximately 8,000 migrants, of whom 1,500 were minors,
swam around or climbed the fence to enter this EU outpost separated from
Spain by the Strait of Gibraltar. Twenty-three migrants died in their attempt
to climb the fence and more than 80 were injured. Many of the migrants
were from Sub-Sahara Africa, some fleeing conflicts in Sudan and Mali, while
others had abandoned their homelands due to the effects of climate change.
Initially praised by EU leaders, the violent repression by Moroccan police
was later criticized by the EU in an attempt to distance itself when the full

extent of the atrocity emerged. The risks and deaths refugees are exposed to in their journeys to seek better lives have become normalized, solutions immobilized, and the issue desensitized to the point of acceptance. Of the 8,000 who attempted to cross into the Spanish enclave, only 133 claims for asylum were submitted from that day with the rest of the migrants being expelled back into the African territory.[2] Nation–state border protectionism from rich Western countries is constantly being upgraded to combat these new migration flows of climate change and economic refugees that openly displays the disparity in the value placed on different human lives. Cheap and exploited in some parts of the world and expensive and protected in other parts—the disparity belies a globally entangled unethical and immoral system at work. But not all refugees suffer from such violent border enforcement. The outpouring of support from the majority of EU member states in face of the 2016 Syrian refugee crisis as a result of the country's civil war saw millions of people migrate to Turkey and the European Union. This initial support, driven by moral and ethical obligations, was largely led by the then German Chancellor Angela Merkel, who saw her country absorb more than one million refugees, though the scheme would later fall victim to the hardening of public and political sentiment.

The effects of climate change on tens of millions of people across the world surfaced well before human impact on the earth was felt. European colonial expansion of the 18th and 19th centuries extended across the globe saw stolen resources and enforced labour on a massive scale to feed Western industrial revolution. This export of resources, subjugation of labour and culture proliferated in the 20th and 21st centuries where consumption and energy needs increased exponentially. It cannot be dismissed that present-day human mobility is an ongoing result of the 300-year history of Western global dominance. If seen in a new light, one could say that the people now forced to leave their homelands to reach rich Western countries for the chance at a better life are in a way seeking to claim back the wealth taken from them. Writing in her book *Border Nation: A Story of Migration*, Leah Cowen argues that 'the demarcation of borders, separating wealth from the sites of wealth creation, has enabled the systematic destruction of the environment. While countries in the Global North...are the top contributors to global temperature increase, countries in sub-Saharan Africa, in particular, are the most dramatically affected by climate change, and least resourced to respond'. Cowen describes the maintenance of division whereby the 'Global North continues to export and outsource production of its goods to the Global South, it also exports its emissions and environmental degradation, meaning that goods and capital can cross borders with ease, while workers are forced to stay put and deal with the damage that has been wreaked on their local environment'.[3] Cowen suggests a new global conflict is at hand and that, in 'the face of the extreme violence of borders—this relentless onslaught which pounds like waves—we must fight back, even when it feels like trying to battle the tide. Rejecting borders and the border nation is not a simple act. We have to dream bigger and bolder than the narrow confines of Home Office check-boxes, bars and barbed wire'.[4]

Rejection and containment to constrain human mobility cannot last and will ultimately overwhelm Western protectorates in the face of the sheer numbers of people affected by climate change. It is worth remembering that less than 200 years ago, during the height of colonialism, the theft of resources and the transportation of enslaved people was an acceptable condition of global European/Western trade. In an article for the *New York Times*, Abrahm Lustgarten charts the rise of South, Central American, and Mexican migrants' journey to reach the United States. Lustgarten begins with a statistic to connect human migration and climate change: 'Today, 1% of the world is a barely liveable hot zone. By 2070, that portion could go up to 19%. Billions of people call this land home. Where will they go?'[5] The answer to this question is being avoided and instead countered by the militarization of borders to force the issue to go away.

The rise of the earth's temperature is leading to fragile regions becoming barren and unsuitable for cultivation, and people living in these regions have two choices: either go south to already over-strained populations in South America, India, southern Africa, and Asia or north to the richer countries of the United States, Canada, and Europe.

> If governments take modest action to reduce climate emissions, about 680,000 climate migrants might move from Central America and Mexico to the United States between now and 2050. If emissions continue unabated, leading to more extreme warming, that number jumps to more than a million people,

predicts Lustgarten. His article contains a graphic showing affected regions marked in a band of red across Central America and the Amazon, North Africa, Middle East, Pakistan, India, South-East Asia, Indonesia, and Northern Australia. On this band where billions of people live, most rely on agriculture for their survival. Scorching heat, drought, acidification, desertification, and water scarcity as a result of climate change means that many leave their homelands for urban centres. 'Millions seek relief, first in big cities, spurring a rapid and increasingly overwhelming urbanization', Lustgarten explains. 'Then they move farther north, pushing the largest number of migrants toward the United States. The projected number of migrants arriving from Central America and Mexico rises to 1.5 million a year by 2050, from about 700,000 a year in 2025'.[6]

Early human migration out of Africa instigated the terrestrial body in movement with geography and terrain and the ability for adaption to changing climates. The idea of place was not part of early nomadic life—a Western concept this would come tens of thousands of years later with the establishment of settlement asserting a sense of belonging through occupation, demarcation, and protection of the land. Nomadic life was dependent on gathering knowledge of the terrain, animal, and plant life. In *Population Genetics and the Migration of Modern Humans (Homo Sapiens)*, Peter de Knijff expresses that nomadic life was genetically fundamental to early humans.

It is very likely that modern humans have always been a migratory species as far as geography and ecology permits. For this reason, the transition by modern humans during the last 10,000 years from a mobile hunter–gatherer life-style towards predominantly stationary farming-based societies should be seen as the exception rather than the rule. One consequence of our mobile past is the non-random global distribution of genetic variation.[7]

Knijff suggests that early humans' genetic makeup was randomly acquired and, as in the case of all species, genetic diversity is key to overall survival through interspecies contact. Knijff points to one likely explanation of early genetic diversity using the example of males where from 'out of Africa they accumulated new Y chromosome variants at different stages of their migration'.[8] Migration, diversity, and connectivity are inter-reliant in attaining gene variation necessary for survival. Given that women contained two X chromosomes, their variation is embodied. X chromosomes have more than 1,000 working genes, whereas Y chromosomes have 100 working genes. The connection between geography and male gene diversity over the stages of early human migration suggests this growth pattern may have been halted when nomadic life for males transitioned to the permanency of settlement and stasis to their surroundings. In his book *Nomads, Empires, States*, Kees van der Pijl equates a different idea of diversity with each in contest with the other. 'On account of their mobility, nomads have been the key antagonists of sedentary empires throughout history'.[9] Pijl's reference to nomads as 'antagonists of sedentary empires' illustrates how mobility, rule, and nationalism radically shifted nomadic life and, in following Knijff's point, would lead to the reduction of male Y chromosome gene diversity. How this mention of human gene diversity relates to climate change will become clearer later in the chapter.

As the foundation of human diversity, nomadic people sustained their lives through their connection to the terrain. The period of time from settlement to the 21st century covering approximately 12,000 years, the remnant gene-pool connectors are now embodied by far fewer population of First Nations peoples whose lives and natural habitats are declining across the world. In Sub-Saharan Africa, the nomadic life of the Maasai is being threatened from sediment farmers who are claiming the lands where their livestock have grazed, threatening thousands of years of environmental adaptation and diversity. In the Brazilian Amazon, indigenous tribes are facing expulsion from their forest dwellings by illegal loggers and miners reducing their genetic pool as a result of the destruction of their environment. In Australia, there were 300 aboriginal nations that occupied the continent before English colonization decimated their culture by forcibly uprooting tens of thousands of years of environmental adaptation and diversity. And in southern Israel's Negev desert, the nomadic Bedouin people are being expelled from their nomadic lands to make way for illegal Israeli settlements resulting in the Bedouins losing their connectivity and identity to their lands and history. The destruction of environments and forced exodus from their lands afflicting the last of the world's nomadic peoples in turn affects the global gene pool of environmental knowledge and adaptation gained over tens of thousands of years.[10]

Terrestrial mobility of nomadic life was key to ecological interaction creating an end-to-end encrypted coexistence with the natural world. Nomadic hunters, gatherers, and herders as a matter of course abided by the geography of a region making the most of a terrain's abundance. Their spatial sphere was inclusive to 'wearing the ecology' of the environments they moved through and the ability for sustainable living through their agility to explore new routes and improvise and adapt to new terrain. The modern-day fight against nomadic people such as the Maasai from sedentary farmers restricting their ability to move in sync with the changing seasons reveals marked differences in adaptation, exposure, and vulnerability to terrain and weather in maintaining their survival. The nomadic life of the terrestrial hunter/gatherer and herdsmen declined as the static occupation of ground and cultivation increased. The transition from nomadic life to static occupation of ground by the cultivator that saw an uprooting of the ecology from the body bears an influence on the genealogy of human life in relation to the natural world now. At a genetic level, the idea of a depleted pool of male Y chromosomes from 100 and female X chromosomes from 1,000 might be one way to explain the decline of human relation with the natural environment. It might also explain how half the world's population (four billion people) living in human-made constructed environment of urban centres are dislocated from the natural world, at the same time as another four billion people living in rural environments face extreme weather turbulence and environmental degradation as a result of climate change. In light of the human evolution to modern humans 200,000 years ago, to their migration out of Africa 100,000 years ago to colonize the world, it is worth asking: what relationship will the future of humanity share with the environment?

Reopening the pathways for human migration might forge a new human evolution and diversify Y and X chromosomes, which is necessary for adaptation to climate change in the 21st century. A new wave of human migration as a result of climate change is already unfolding, though it remains systematically suppressed. One major problem is the territorial protections of richer nations in repelling human mobility. Territorial borders are designed to stop people from crossing over, whether in the form of the physical fences and checkpoints on land or the invisible boundaries demarcating nation-state sea borders. Where nomadic life declined with the establishment of settlement and agriculture, the universality of climate change is challenging the system of belonging, national identity, and citizenship. As climate change intensity increases, affecting more regions and more people across the world, national identity and citizenship enhanced by border protectionism must be considered outmoded and counterproductive. The dichotomy confronting global sustainability lies in the atmospheric and non-territorial oscillations of weather turbulence to on-the-ground divisions of nation-states. In her chapter, 'Refugees—Performing Distinction: Paradoxical Positionings of the Displaced' from the book *Geographies of Mobilities: Practices, Spaces, Subjects*, Alison Mountz argues that for refugees, 'status is doubly bound to nation-states. They are displaced and ask for protection from a state where they must prove well-founded fear of persecution and lack of protection by the state where they hold or once held citizenship'.[11]

Ideological, sexual, religious, and gender persecution of refugees is well documented and accepted by liberal nation-states willing to grant asylum. Far less so are climate change refugees who suffer a different set of threats not of their making, which uproots their lives and enforces their migration. Indisputable and evident in the hundreds of millions of people affected by climate change throughout the world, their fate is far more difficult for the global community to manage given the scale. Rejection has become a sure-fire way in the short term of managing what will become unmanageable in the long term. The border regime will eventually collapse in the face of the sheer volume of people seeking refuge from climate change. Given the speed of technological innovation that afforded modern migrants crossing the earth in the 19th, 20th, and into the 21st centuries, most of humanity is bound to territorial borders of national divisions. Differences in the perceptions of race, much as a passport identifies its holder's nationality, are based on certain restrictions depending on which country one is from. In her book *Resident Foreigners*, Donatella di Cesare argues that human mobility was altered at the point where national identity forged the claim for a territorial dominion.

> In the past there were no lack of forms of movement, from nomadism to military conquests, from invasions to bold and adventurous journeys, up to and including the first real efforts to found colonies. All these forms of movement were dominated by the collectivity: it was the group that moved, as it sought to establish or widen its dominion over some territory.[12]

When a group of people claim a territory as their dominion, anyone outside of their claim is regarded as foreign and any trouble from within is punishable by banishment, an idea that still holds today.

Exclusion and expulsion are still the basic foundations of nation-states' authority to enforce their territorial claims over incoming migrants. 'Exiles, the stateless, the proscribed, fugitives, refugees, displaced persons, asylum-seekers, emigrants, nomads, "illegals"—the list of more or less discriminatory terms varies across different languages and widens every time that new elements emerge in the multifaceted world of migration', Donatella explains.[13] These same recriminatory titles are being applied to climate change victims who by default have become incriminated by the excesses of consumption and pollution caused not by them but by rich Western countries, the same countries that are restricting the victims' mobility. To incriminate people because of their freedom of movement should be regarded as immoral under human rights, but to go further and incriminate climate refugees only amplifies the immorality. Rich nation-states base their rejection of climate refugees as their right but, as we have seen, there is no right or wrong or divisions when it comes to effects of global climate change. In an age of extreme climate instability, nation-states' peddling of nationalism, protectionism, and fear onto their citizens at the expense of the most vulnerable people is a global human rights issue, as such rights are only granted to the enforcer.[14] In *Border Nation*, Leah Cowen observes the clash between the hard border mentality deployed by Western countries and the soft embodied vulnerability of the migrant. 'Work to amend immigration laws and policies is important insofar as it improves people's lives in the here and now—but we must carefully consider the ways in which "reforming" an inherently abusive border regime makes it harder to destroy altogether'. Cowen considers the hard border mentality through the analogy of a balloon: 'Just as air inside a balloon would never advocate for bursting the skin that holds it, states as we know them will never support steps towards their own eradication'.[15]

One major difference between early human settlement and present-day territorial boundaries of nation-states is the accessibility of information provided by global telecommunications. Images of wealthy Western lifestyles beamed onto smartphones around the world reveal the lifestyle gap between the rich and the poor but share none of the discriminatory boundaries that humans experience in the physical world. It is only natural that people affected by climate change, poverty, and a lack of opportunities should seek refuge in rich, least-affected countries where opportunities are abundant. The protectionisms and collisions that present-day climate change migrants are confronting are disingenuous to the messages of global telecommunications accessible to anyone with a smartphone. Visually desirable yet unreachable, images of wealthy carefree lifestyles on global media platforms point to the abandonment of humanity's moral code to the struggling reality of the majority of people, marking a dark turn in human history. As the numbers of climate change victims rise, new solutions have to be found to sustain, not reject, their lives. If no new solutions are found, the obvious outcome to this sanctioned inequality and discrimination is human conflict, violence, and war. Given humanity's history to solve conflicts through war, a global conflict is fast emerging on a scale previously unprecedented in human history. Managing global human mobility in tandem with programmes of regional adaptation to changing climatic conditions is the most effectual way in which humanity can lead itself in responding to the effects of climate change. Territorial boundaries and protectionisms are fast becoming redundant divisions in the face of the scale of environmental devastation that is affecting hundreds of millions of people across the world.

Global sustainability as objective and practice remains limited in its approach to solving the human cost to reclaim climate security. What then, we might ask, is the intention of sustainability? Since climate change has no boundaries, sustainability must also entail a systematic revolution of the world order. The present order of political, capital, cultural, racial, gender, and religious divisions that are geographically rooted between the so-called Global North and Global South calls for a new world charter to be drawn up and designed on the basis of human equality in achieving global sustainability. The United Nations has such a charter but its structure of membership is prone to the manoeuvring of countries, which makes this international body ineffectual in the face of the geopolitical forces to which it is tied. With the earth struggling from the effects of human impact including deforestation, land degradation, desertification, water scarcity, drought, and flooding, much of it a consequence of the excesses of the rich minority of the world's population but prevalent in the everyday lives of billions of people of poor countries, such proposals for a new global order have to be negotiated as a pre-eminent condition to reaching global sustainability. Yet, chipping away at global inequality caused by climate change will not be enough to secure the future of human and the earth coexistence. The ancient terrestrial nomad 'wearing their ecology' as they crossed geographical regions left minimal impact of their presence. From early settlement through to kingdoms and colonialization, nomadic tribes were viewed as disruptive to the progress of static communities. This 'disruption' is still evident today such as lives of the Maasai people in Tanzania where expulsion from their grazing lands was recently undertaken in favour of luxury safari hunting and elite tourism. The infinite geographical space and knowledge nomads carried within their bodies was expunged and subsumed to the definitive boundaries and the plundering of resources through European expansion. The defining point of land acquisition, first conquered then delineated by lines on paper and positions on maps as proof of ownership, meant that the original inhabitants were both displaced from their lands and held to the ravages of their conquerors. The ongoing clearing of forests and the decimation of whole regions through mining continues as in the Amazon, the Congo Basin, South-East Asia, and South America. Sedentary occupation and cultivation of land removed nomadic life as the pre-eminent way of inhabiting the earth. To continue to remove the remaining nomadic peoples of the world from their lands will mean that their environmental knowledge and adaptation will become extinct and this overall will have profound effects on humanity's ability to adapt to the effects of climate change.

Globally supported terrestrial mobility in the 21st century recreates the ability for humanity to connect with the natural world, gather environmental knowledge, undertake indigenous land care, and build alternative living structures (societal and physical) to better manage the earth's ecologies. Through a combination of mobility and adaptation, ideology and practice, meeting the global effects of climate change is then established in all parts of the world not just the most affected regions and populations. The present quandary facing the world in an era of extreme weather turbulence is not one to be drawn into territorial confrontation. In an age where products shipped across the world have more rights concerning freedom of movement than the lives of the most vulnerable people, this speaks of the disparity (people and products) the world has been built on. The nomadic hunters, gatherers, and herders wore their ecology as the conduit to the environmental conditions of regions, weather patterns, and the availability of resources. Their foreignness to the concept of land ownership and their mobility to move and adapt to different climates averted their exposure to damaging climatic conditions. The establishment of settlement vastly improved human survival and population explosion but it has come at the cost of selective mobility, inequitable power, a capitalized global marketplace, and nation-state territorial protectionism. Where the nomad did not build fences to demarcate their land, it has become clear that the dominant sedentary habitation of the earth today has exposed human vulnerability to climate change. New avenues must be engaged and structures established to support new pathways for human migration across the earth as one of many ways to secure people's survival. Contemporary forms of temporal occupation of ground and forming of societies connecting city and rural regions will help to close the gaps and diversify human capacity for adaptation. To build resilience in the face of climate change is to respond to the borderless impact of global warming by removing restrictions on human mobility for a new supportive system. People affected by extreme weather events must be given the ability to determine their lives and seek new opportunities for survival especially so when climate change is not of their making.

Settling—Restraining Fields

The establishment of settlement curtailed nomadic life that characterized human habitation on the earth for hundreds of thousands of years. Dwelling on open ground came with the confidence of building-in protection that settlers could claim dominions beyond the walls of their settlement. No longer were humans subjected to the insecurity of sourcing food via hunting–gathering; food supply was now stabilized through the cultivation of crops and the domestication of animals. Settlement changed humanity's relationship with the natural environment by enabling a psychological separation between people and their surroundings. 'It was men rather than women who were particularly prone to bouts of timidity', Joanna Bourke tells us in *Fear: A Cultural History*. 'Fear could ascend for irrational reasons; it could subside just as impulsively'.[16] This new construct of environmental separation between humans and physical geography into fear and threat ushered in the spatial trajectory for the forming of national identity, civil society, and citizenship. The imaginative–hallucinogenic effect that timidity and fear brought to settlement continues in various forms in the urban centres of today but where timidity and fear are internalized between people and the city itself. Dispensing with nomadic life for the permanent occupation of ground altered the settler's natural balance with the environment. Now parcelled into provisions of food production, the natural environment need not be 'natural' for human survival but 'cultivated', shifting from insecurity to domination of their surroundings.

From settlement to the industrial speed of production and consumption up to the 21st century, we have seen an incremental withdrawal of human ecology with the natural environment in favour of shaping it according to human needs. This is important in understanding the threads of human displacement and nowhere is this more acute than in the mega settlements of today's cities, the ribbons of highways crossing landscapes in cars set to cruise control and jetting on streams of compressed air 10,000 metres above the earth in the sealed compartments of aeroplanes formulating a superolateral connection to the natural world. Enhanced with the time–space compression and technological inventions, humanity, especially urban dwellers, imposed a form of self-forgetting in terms of the devastation on the natural environment caused by the excesses of continuous growth. After centuries of building-up gross resource extraction and environmental controls, humans now fear the planetary chaos of their impact. What has emerged in urban centres as distinct from rural regions is a psychogeography of human disorder to the natural world. Referencing the French Situationist of the Lettrist Group Guy Debord, Merlin Coverley in *Psychogeography* notes that 'in broad terms, psychogeography is, as the name suggests, the point at which psychology and geography collide, a means of exploring the behavioural impact of urban place'.[17] From settlement to citadel, nation-state to colonialization, to the industrial revolution and globalization, it was obvious that humanity's connections with the earth would eventually become unsustainable. The devastating effects of climate change on the natural world from the burning of fossil fuels polluting the atmosphere to deforestation, mining, and toxic pollution of groundwater, rivers, and lakes have altered the natural balance of the earth, resulting in catastrophic disruption.

Humanity's environmental psychogeography is not a straightforward or equal one. In rural regions where the other half of the global population reside, farmers, indigenous peoples, remaining nomadic herders, hunters, and gatherers, forest dwellers, and islanders who have to some degree retained their 'natural' balance with their environments are now colliding with the effects of climate change. Where there was once harmony with their surroundings, now discord resides. In dense urban centres, psychogeography collision had been in existence since settlement. To unscramble the urban psychogeography of environmental disconnection, cities and their inhabitants have to be reprogrammed, decentred, and naturalized and bridges built between them and rural populations. As long as the divide between urban and rural populations remains, the capacity to fully grasp climate change and understand the lives of those most affected will continue to disrupt the push for global sustainability.

As the global urban population is set to increase from the present 50% to 70% by 2050, human impact will continue to de-settle nature, and catastrophic weather turbulence will continue to devastate human and all other life on the earth.

The original fear and timidity that men afflicted on nature, women, and society would also become one of the hallmarks driving the repression of First Nations peoples through European expansionism. Colonialists violently submitted people under their rule to cultural decimation, excommunicating networks, removing natural care of the land to set the course to plunder resources and finally redraw conquered lands into parcels of foreign ownership. Such are the histories of North, Central, and South America, Africa, Asia, Australia, and Oceania.[18] In *The Wretched of the Earth*, published in 1963, Frantz Fanon references the violence of colonialism and the violence that ensued during decolonialisation centuries later constituted the same form of violence: the sanctioned repression of indigenous peoples in the case of the former, and the eradication of white supremacy in regaining independence in the case of the latter. 'You are rich because you are white, you are white because you are rich', he writes.

> The violence which governed the ordering of the colonial world which tirelessly punctuated the destruction of the indigenous social fabric, and demolished unchecked the systems of reference of the country's economy, lifestyles, and modes of dress, this same violence will be vindicated and appropriated when, taking history into their own hands, the colonized swarm into the forbidden cities.

In the time Fanon's book was written, the march for independence in many former colonial countries was unfolding. 'To blow the colonial world to smithereens is henceforth a clear image within the grasp and imagination of every colonized' subject, Fanon continues. 'To dislocate the colonial world does not mean that once the borders have been eliminated there will be a right of way between the two sectors. To destroy the colonial world means nothing less than demolishing the colonist's sector, burying it deep within the earth or banishing it from the territory'.[19]

Across the world on the African, South and North American, Asian, and Australian continents, independence movements were freeing themselves from the yoke of European colonialism but its systems of inequitable exchange remained. 'We are witness to the mobilization of a people who now have to work themselves to exhaustion while a contemptuous and bloated Europe looks on', Fanon writes, citing the long hardship of independence that African and South American countries underwent in 1960s and 1970s. 'Capitalist exploitation, the cartels and monopolies, are the enemies of the underdeveloped countries'—an observation that remains evident today in many post-independence countries.[20] In the preface to *The Wretched of the Earth*, Jean-Paul Sartre outlines with cynical reality the history of colonialism:

Not so long ago the Earth numbered 2 billion inhabitants, i.e., 500 million men and 1.5 billion 'natives'. The first possessed the world, the others borrowed it. In between, an array of corrupt petty kings, feudal lords, and a fake, fabricated bourgeoisie served as go-betweens. In the colonies, truth displayed its nakedness; the metropolises preferred it clothed; they had to get the 'natives' to love them.[21]

Globalization was the attempt to get the world to 'equally' participate in the capital of global trade but the reality of this market system has only reached 30% of the world's population. The practice of globalization started well before the concept came into being in the second half of the 20th century. The Age of Exploration and the colonial invasions that followed spread the European perspective towards nature and the establishment of global inequity. Early circumnavigation of the oceans brought a measurable calculation between distance and time across the globe. Where the physical world did not alter in dimension, ocean navigation inadvertently shrunk the world to a course plotted on a map. Central to acquiring distance and location on the high seas was John Harrison's invention of the chronometer bringing longitude to pin-point accuracy in which comprehensive and reliable trade routes could be established.[22] Metaphorically shrinking the world to a hand-held mechanical device and charting shipping routes for efficient commercial trade brought the riches of the world under European control. European navigation and exploration could have been an opportunity to share cultures and trade equally, which would have resulted in a very different and more equitable global order as well as a very different environmental history. Instead, what prevailed was the white supremacist European mentality, military technology, and ignorance that established the cross-continental trade ruled by invasion, dispossession, enslavement, and killing. European colonialization also brought Christianity and the concept of divinity of the human soul to claim control over the life and death of all living things. This religious belief was aligned with the colonialist perspective on First Nations peoples, their bodies, and their land. Over the following centuries the European perspective on the natural environment was exported to these new lands—forests were cleared, minerals mined, animals, pests, and diseases introduced. From the Age of Exploration through to the 20th century, indigenous knowledges of the land were summarily stripped of their worth in favour of the fortunes that the land could provide. European expansion converted environments into sites of profit leading a conversion of values that has led to climate change today. What started out as 'European discovery' across the world was not the expansion of human relations between peoples but subjugation, cultural decimation, and expulsion. Globalization began with the Age of Exploration as the effects of climate change began with colonialism.

The global impact of colonialism is measurable by the catastrophic effects it exerted on people and their lands through to the impact of climate change now. The Spanish conquests in Mexico, Central and South America laid waste to indigenous cultures and after centuries of repression they are now finding political and militant voices to recover their culture and lands. Success in reversing the effects of colonialism in one country and continent is counterbalanced by the silencing in another where institutional and corporate corruption prevails over indigenous peoples' rights to take back their lands. Indigenous Peruvians, Bolivians, and Paraguayans are some of the South American peoples charting an independent course for the return of their lands. In America and Canada, the 19th-century forced expulsion of native peoples from their lands and incarceration on reservations has, in the 21st century, led to a strengthening and resisting of the encroachments of mining and fossil fuel corporations seeking to further exploit their lands. In *Reclaiming Indigenous Voice and Vision*, human rights lawyer, James (Sakej) Youngblood Henderson of the Chickasaw Nation and Cheyenne Tribe Oklahoma writes: 'We must continue to see the organization of life in terms of the Indigenous knowledge about living in balance with an ecology. We must use our traditional knowledge and heritage to force a paradigm shift on the modernist view of society, self, and nature'.[23] One of the hallmarks of colonialism was to remove First Nations peoples' connectivity to their land. Europeans gave little respect to their knowledge of ecological sustainability. 'No ecology', Henderson writes, 'no culture, no people, and no psyche remains untarnished. Oppression was everywhere, and so was the technology of social control and death'.[24]

Moving from North American Indian cultural decimation, the 300 indigenous nations that covered the Australian continent in a porous seam between tribes before colonial invasion are now regaining their lands and culture. Not recognized as the original owners of the country they have occupied for 60,000 years—the longest continuous occupation by humans—their knowledge of the terrain is now being enlisted in managing the land to reduce bushfires and the decimation of native animals. On the African continent, many former colonial countries gained independence from their colonial rulers in the latter half of the 20th century. Up until then African countries suffered immensely from their European colonial occupiers, with mining and oil companies extracting their resources unabated.[25] In many other post-independent colonial countries such as in Central and South America, India, Asia, and Oceania, the majority of their people still live with the remnants of colonialism with fewer opportunities to better their lives than those in Western societies. The destruction of indigenous cultural heritage through military force, enslavement, and religion cannot be underestimated in the context of global sustainability for the 21st century. Severing indigenous peoples' links to their natural environments has resulted in irreversible ecological destruction in many parts of the world, severely affecting their continued survival.

In his book *To Cook a Continent: Destructive Extraction and the Climate Crisis in Africa*, Nnimmo Bassey describes Africa's history, noting 'how the invaders conned the continent with a bible in one hand and a musket in the other'. Paraphrasing the South African Archbishop Desmond Tutu, Bassey writes: 'when the missionaries came to Africa, they had the bible and we had the land, but they asked us to close our eyes and pray, and when we opened them, we realized they now had the land while we had plenty of bibles'.[26] The focus of Bassey's book is the inequality of colonial history to present-day corporate imperialism most notably by European, American, Australian, and Chinese mining, oil and gas companies on the African, American, and Asian continents. 'Conquest meant division and disunity across the continent', he writes.

> Conquest meant the splintering of nations and kingdoms into different blocs; it meant the amalgamation of disparate units into new wholes in a tensile state that promised no peace. Conquest and division laid prostrate vast civilizations in Africa, the America's and Asia, siphoning off resources to fuel the industrial revolution in Europe.[27]

Colonialism split the world into two distinctive camps: plunderer and the plundered. The Age of Exploration which had made it possible cast the world into this division by creating global access by one part of the world to rule over the 'discovered' lands, setting up the system of global racial suppression and inequality. Centuries later, the world is still divided into two distinct opposites, now called the 'Global North' and the 'Global South'. If we believe that the world is getting better at addressing the histories of colonial suppression, we don't have to look far to see these same suppressions at work—all of which obstruct the implementation of sustainability that the global community is trying to reach so as to combat the effects of climate change.

Reconciling colonial histories is just as fundamental to achieving global sustainability as is the limiting of resource extraction, fossil fuel energy production, chemical processing, and industrial pollutants in the air. How the effects of climate change are disparately and differentially experienced by the world's eight billion people, split between urban and rural populations, needs to be better understood and resolved. It is not possible to reach global sustainability if one half of the world's population keeps on increasing product consumption, while the other half suffers most from the effects of climate change. Reaching sustainability is also not viable when global corporations continue to profit from the use of fossil fuels and governments continue to advocate continuous growth. While it would appear that urban populations, which are by far the larger emitters of CO_2, have the greater possibility in combatting climate change, collectively they have far less direct experience of its effects. Within the sealed environments that protect them from the outside elements, the world's urban populations are removed from the daily struggle that rural populations experience, such as devastating weather events that directly imperil their survival.

The day-to-day survival of both urban and rural populations across the world is increasingly becoming normalized as a way of living in the 21st century. There are approximately 3.6 billion people exposed to climate change and 2 billion of those people face water and food insecurity on a daily basis where women and girls walk miles to fetch clean drinking water, live with open sewers, and die from preventable diseases. This is of course different from urban populations where water appears effortlessly from taps, cool air from air conditioning ducts, heating via continuous supply of electricity, sanitation controlled via a network of underground pipes, and food filling supermarket shelves. In the first two decades of this century, climate change reports have saturated TV screens showing in real time the human impact of climate change. Where a sense of despair, depression, and action occupy people's minds concerning climate change, desperation, and the threat of survival is the reality for the majority of the world's population. The impact of climate change and the ongoing residual effects of the colonial era are further dividing the world in numbness, anxiety, and increasing acceptance.

Various statistics indicate that the global exodus of hundreds of millions of people leaving their homelands will bring about the new mass migration across the world. Global telecommunications reaching the smartphones of billions of people have in effect extended the realization of human exodus from places unable to support human life to places that offer opportunities for a better life. The two billion people around the world suffering extreme water and food insecurity cannot be overlooked by richer nations entrenched in positions of 'us and them'. A continuation of this mentality can only lead to mass human conflict. If we follow the historical accounts of slave revolts against masters during the Roman Empire, rural uprisings of the poor against their landlords in medieval Europe, repressed minorities against majorities and later independence movements from colonial rulers, the defeat of apartheid in South Africa, the struggles of the working class against the barons of industry, we see that the pendulum swings between downfall and resurrection. Sustainability, it could be argued, moves in the same way from what is needed today to secure an environmentally stable future tomorrow against the ever-increasing consumption of the world's resources. Sustainability is currently being guided by the maintenance of power and protection just as men peddled fear and timidity over women and society in the establishment of settlement. The perpetuation of such histories and systems leads not only to the whitewashing of sustainability but also to the downfall of humanity.

Expelling indigenous peoples from their lands, territorializing geography by boundaries and not by cultural associations, and making off with the natural resources without any recompense to the original owners reveals that sustainability has little chance of becoming a truly global movement if these histories and practices are not addressed. The establishment of settlement saw the systematic destruction of nomadic life. Colonialism through to globalization saw the systematic decimation of indigenous cultures to the benefit of the homogenized societies of the invaders and industrialists. The new division of urban and rural populations defined by protection, exposure, and vulnerability to climate change is the new frontier of the world that needs to be reckoned with. The restitution of nomadic life is one part of creating a diversity of responses to living with climate change; the other is that of reconciling the histories of the past. Where the stasis of urban life in human-made constructed environments has replaced the natural world, nomadic life offers the ability to (re-)gather ecological knowledge for adaptation in radically changing environments. How to relieve countries of their fear and timidity to the mobility of people seeking to evolve as the earth evolves in response to climate change is a major obstacle to sustainability. Sustainability cannot be designed as a suitable system aimed at maintaining the lives of the privileged. It must be life-practiced; it must allow new forms of habitation to emerge. It is by the dexterity and flexibility of its implementation that future generations will have multiple ways of being in the world than the limitations of today's inequitable system.

Territorial—Digital Accessibility

A viewing of the world map shows straight and wiggly lines denoting the separation from one country to the next. Between these lines are other markings denoting geographical features such as rivers, valleys, mountain ranges, forests, and deserts that characterize a region, country, and continent. Once the sole enterprise of mapping was to render the world in two dimensions, the illustration has now moved from static paper to the spatial mobility of mobile phones, turning once incalculable distances into GPS satellite tracking measurements. As touched on in the previous section but warranting further mention here, global communication now shows the world moving in real-time visual graphics of smartphones used by the walker to track their movements over ground and in sync with the inner geological shifts and outer axial spinning of the earth's inter-planetary orbit. Digital representation of the world moves seamlessly across the physical barriers of borders and fences, in contrast to migrant bodies with their smartphone devices in hand being repelled at the crossings. Having used the phone to navigate terrain they have walked, crossing rivers, deserts, mountains, and seas, the accessibility of the world as digital display does not convey the realities of its divisions. Whilst technology has given incalculable freedom of information, real-time display of terrain, weather forecasts, and alerts of danger, it does not convey the extruded constructions of borders.

Well before the technologies of digital and satellite communications, the world was drawn in geographical features of land mass and the oceans in measurable dimensions. The art of cartography represented the world in two-dimensional space drawn on paper that could be rolled, transported, and unrolled in any location to a marked position on the map to the physical terrain of the viewer. As will be further explored in Chapter 5, maps transformed human comprehension of the earth to be surveyed and claimed. As radical displays of space, maps tabled land and oceans and with the point of the compass and sextant in relation to the angle of the sun to the horizon to determine locations on land and sea via longitudinal and latitudinal dimensions. Long before these instruments and use of geometry, early human representations of terrain and location were evident in cave paintings and far later in Mesopotamian, Egyptian, Chinese, Greek, and Roman cartography to enable a centralized overview of existing domains and the possibility of locating new dominions for expansion. From concentric lines denoting the rise and fall of terrain to jagged peaks in hatched and sharp lines, maps were a crucial tool in planning new routes across land and sea.

In the 21st century, the digitalized rendering of the world on mobile devices in the hands of billions of people has continued to convey the sense of a centralized position to the world. The image of the world can be zoomed in and out, locations identified, journeys planned, and distances calculated. People can now walk with the world in their hand and when they move, the image of the world moves with them. What these relatively new centralized images of the world do not convey, however, are the differences between the visually accessible world to the realities of selective accessibility of the human body. For the majority of the world's population, especially those people from the so-called Global South, the truth of holding the world in one's hand was the reality that their position was in fact not centred but peripheral. The contradiction of visually holding the image of the world and simultaneously being selectively rejected from it forms an in-hand yet out-of-reach world for billions of people. This contradiction is further enhanced with the accessibility to images that show the disparity between people in their standards of living throughout the world. People's lifestyles in far-off countries could be viewed in a kaleidoscope of images that meld and divide the world in a seamless flow of images of the lifestyles of the rich as against the lives of the poor. Desire, relief, hope, and opportunity are displayed in the condensed dimensions of images in the hand where both proximity and immeasurable distance and the experience of lives lived are as distinct as they are discriminatory.

Throughout history, technological innovation has shaped human perception and interaction with the earth. In his book *Home Territories: Media, Mobility and Identity*, David Morley reveals that 'various forms of electronic communication, from the telegraph onwards, have had the effect of creating a greater sense of such coevalness, in so far as they produce a wider sense of a shared present across geographical space through a sense of broadcast liveness as a form of fellowship'.[28] Imaging the world in digital accessibility gives the impression of a world equally shared, as if we are all somehow joined; for the same image of place or location on the map is accessible to anyone in any part of the world. Yet, as Morley points out, the world is displayed differently to highlight its separation where 'the typical mode of the representation of these distant others to us via the mass media, the people of this faraway world are often closely associated with a world of trouble—of social and natural disasters, murders, epidemics and the breakdown of social order'.[29] In Stephen Castle and Mark J. Miller's *The Age of Migration: International*

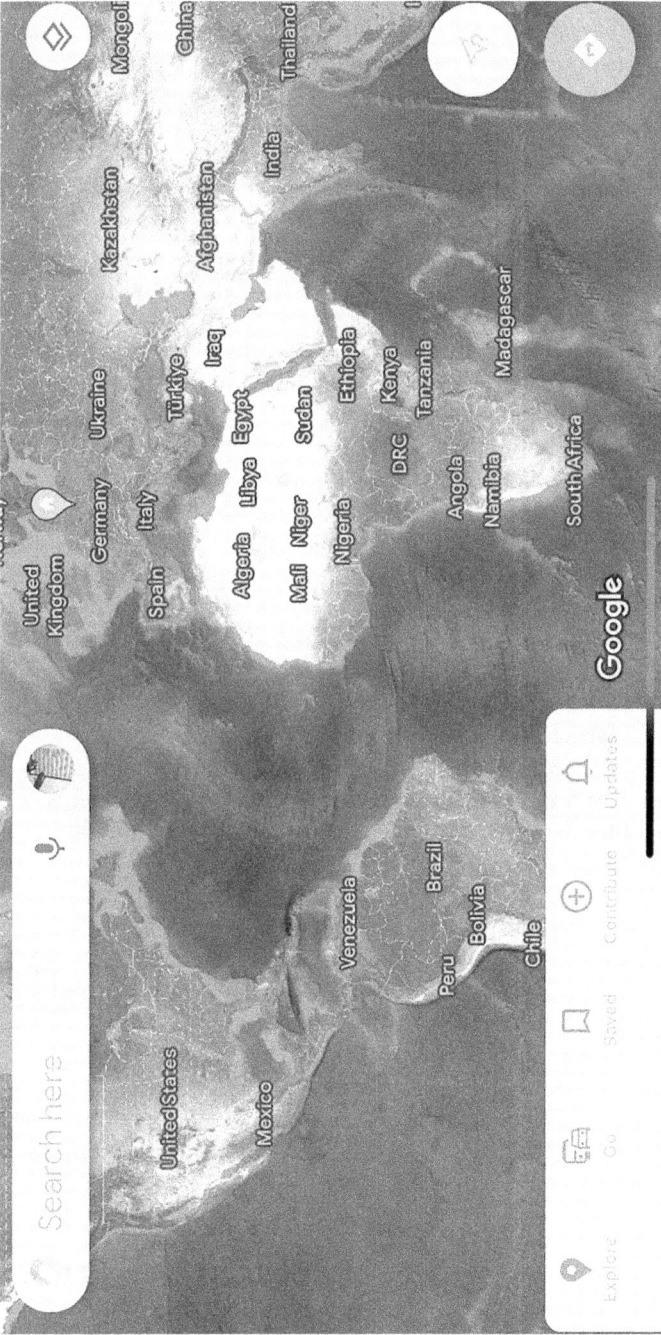

Figure 2.1 Smartphone map of the world.

Source: Image by author.

Population Movements in the Modern World, the authors describe how the 'North-South gap'—that is, the 'differentials in life expectancy, demography, economic structure, social condition and political stability between the industrial democracies and most of the rest of the world—looms as a major barrier to the creation of a peaceful and prosperous global society'.[30] Again, digital accessibility and reality enforces a world not at peace with itself and unable to uniformly prosper. It is an argument that serves the Western world well in keeping what it has by keeping away those who want a share of it, which Willem van Schendel argues is the contrast 'between *fixity* and *motion*'. In *Illicit Flows and Criminal Things: States, Borders, and the Other Side of Globalization*, he writes: 'Borders are presented as spatially rooted, solid, and durable entities, undeniable lines inscribed in the landscape, only to be moved very occasionally and in exceptional circumstances such as war or state disintegration. Illegal flows, on the other hand, are presented as highly mobile, capricious, and unpredictable, improvising new routes as they move across space'.[31] The digital representation of maps and images accessible to hand-held devices explicitly render the world's divisions without showing the separations, and sandwiched between these divisions lie the desires and protectionisms of the vulnerable and the privileged.

Digital presentations of space, regions, cities, and the globe have radically shifted many people's perception and understanding of the world. Both pull and push, attraction and repulsion, desire and differences are animated with each swipe of an image and movement of a digitalized map. For one group of people, it gives possibilities for places to 'visit' and new experiences to have; for another, far larger group, it presents the world as a place of 'viewing' but one from which the physical self is excluded. Images and maps on mobile devices bring a real and visible world in unequal representations of desire and inaccessibility. It is a world of virtual geographies of unreal places that exist both within and outside human comprehension. For people living in the 'Global North', the 'Global South' is a collective imagery of struggle and poverty as much as an affordable adventure. The south is the 'other' to the predominately white people of the north. For centuries, the colour of skin and place within the world created the systematic racial, cultural, and economic exploitation between north and south. Across the globe, humans share in this digital access of an ephemeral and virtual world. 'The ephemerality and accelerated rates of exchange that electronic networks facilitate influence, in turn, how we understand the materiality or immateriality of digital technologies', states Jennifer Gabrys in her book, *Digital Rubbish: A Natural History of Electronics*:

> Dematerialization can further constitute a way of making technologies seem even more operational and effective. The sense of dematerialization, in this case, may emerge through the speed of exchange and space of the interface, which foreground the transfer of signals and light in place of the supports of chemicals, metals, plastic, and labor. Here is a process of

erasure—as well as a process of substitution that works toward a new performativity in the form of accelerated exchange and output. Such erasure unfolds through the speed of electronic networks but also through the apparent immateriality of the software that influences the 'functionality' of those networks.[32]

In terms of migration and global accessibility, digital technology has plunged the world into a hybrid mix of the virtual and the real. At one end digital access opens the world to expose its differences and desires; it has radically changed the knowledge of most people across the world who have access to it and it has revealed how people live differently. It shows the world moving in real time with the walker of a world that is 'theirs' if only pictorially. It has shifted the ground of human migration as it has exposed the different values placed on human life. To some people, the images and digitalized maps of the world are an exercise in the cruelty of this value system; to others, it is freedom of access to anywhere on the map. For sure, no-one escapes from what the digital world represents. For the time being, this arrangement between denial and accessibility to the world is being sustained by enforced separation between the north and south. The length of time that this can be tolerated is getting shorter by the day. The cataclysmic effects of climate change are becoming ever greater and its destructive effects are resulting in a global movement of people following both the 'lies' in the accessibility their smartphones present and the desires contained in the images they see in their hands.

To summarize this chapter is to better understand the human cost of climate change: the physical borders of expulsion and the digital attractions of a visual and portable, mobile world. In *Stuck in Libya: No Escape from Hell* (2021), the French documentary filmmaker, Sara Creta exposed the dire conditions African migrants face in their attempts to reach Europe. One of the migrants interviewed detailed the harrowing treatment in detention camps in Tripoli and was asked why he was putting his life at risk. He answered with three words: 'freedom, justice and peace'. Validating his human right to seek opportunity in the hope of securing a better life is not the same human right that the governments of Europe, North America, and Australia and elsewhere are willing to give. The detention and exploitation of migrants at migrant crossing points clearly flouts the human rights convention to which countries have signed up. Repelling migrants on the basis of the colour of their skin, religious beliefs, customs, and culture, and highlighting a perceived threat that they pose to Western societies does not speak to the issue of climate change but is a continuity of the colonial perspective. To make climate change racial is abhorrent, yet it follows in the footsteps of how racial discrimination and inequality have been institutionalized over the course of centuries. The fact that goods in shipping containers enjoy more rights to mobility than climate change refugees poses a significant test to the resolve of humanity and its direction of travel in the 21st century.

Notes

1 For more on early human migration, see the *Smithsonian Magazine* article: www.smithsonianmag.com/history/the-great-human-migration-13561/

2 In a recent statement from October 2020, a UN representative reported that when talking about climate change, you are also talking about human mobility and migration, and that to address climate change is to address the fallout on people directly and forced into migrating from their affected homelands. Daily TV news coverage of Sub-Saharan Africans shows them risking their lives at the hands of Libyan people smugglers to cross the Mediterranean Sea in overcrowded and dilapidated small boats that capsize, resulting in many drowning in their attempt to reach Europe, most notably Spain and Italy. We also see coverage of Central and South American people making their way through Mexico to reach the United States border and cross into the United States who are either caught and detained in detention centers, drown in the Rio Grande or die from heat exposure of the Arizona desert. Yet, the least reported and biggest migration is the 130,000 people annually from the Horn of Africa into terrorized Somalia, across the Gulf of Aden into war-torn Yemen and on to Saudi Arabia often leading to mistreatment, being forced into slave labour or expelled.

3 See Leah Cowen, *Border Nation: A Story of Migration*. London: Pluto Press, 2021, p. 136.

4 Ibid., p. 126.

5 Abrahm Lustgarten's article appeared in the *New York Times Magazine* in a partnership with *ProPublica* on 23 July 2020: www.nytimes.com/interactive/2020/07/23/magazine/climate-migration.html

6 Ibid.

7 See Jan Lucassen, Leo Lucassen, and Patrick Manning eds., *Migration History in World History*. Boston: Brill, 2010, p. 39.

8 Ibid., p. 44.

9 See Kees van der Pijl, *Nomads, Empires, States*. London: Pluto Press, 2007, p. 62.

10 The nomadic peoples mentioned differ widely in their relationship to our understanding of nomadic life. Western anthropology defines hunter/gatherers, for example, Australian Aboriginals, as distinct from pastoral nomadic people, for example, Maasai people, defined as herdsmen cultivating livestock and peripatetic nomadic people who trade along nomadic routes and settlements. Dwelling, indwelling, and nomadic ways of life might separate these people's relations to their natural habitat but all are connected through their mobility and understanding of their geography.

11 See Alison Mountz. 'Refugees—Performing Distinction: Paradoxical Positionings of the Displaced'. In *Geographies of Mobilities: Practices, Spaces, Subjects*, edited by Tim Cresswell and Peter Meriman. Farnham: Ashgate, 2011, p. 257.

12 Donatella Di Cesare. *Resident Foreigners: A Philosophy of Migration*, translated by David Broder. Cambridge: Polity Press, 2020, p. 25.

13 Ibid., p. 103.

14 Di Cesare refers to John Steinbeck's compassionate story of migration during the Great Depression in America that created millions of internally displaced people.

> At first, they spark compassion, then disgust, and finally only hatred. Expelled from the civil collective, without being able to put up any resistance other than

their blind rage, their dark affliction, they resign themselves to taking leave first from the status of citizens and then from the human condition itself. The others consider them 'domestic barbarians', nomads doomed to the 'state of nature' because they have not grasped the wellbeing of settled existence.

(Di Cesare, *Resident Foreigners*, p. 214)

15 Cowen, *Border Nation*, p. 6.
16 'Fear', Bourke writes, 'could ascend for irrational reasons; it could subside just as impulsively', and would later be projected by men onto women. As 'it was men rather than women who were particularly prone to bouts of timidity' concerning the unknown beyond the walls of settlement, they transferred their fear onto women and in doing so sought to control the female population. The subjugation of women is the result of the fear of men—much of it unknown. See Joanna Bourke. *Fear: A Cultural History*. Emeryville, CA: Shoemaker and Hoard, 2006, pp. 3–4.
17 Coverley further highlights Debord's association with psychogeography, writing:

The origins of the term are less obscure and can be traced back to Paris in the 1950s and the Lettrist Group, a forerunner of the Situationist International. Under the stewardship of Guy Debord, psychogeography became a tool in an attempt to transform urban life, first for aesthetic purposes but later for increasingly political ends. Debord's oft-repeated 'definition' of psychogeography describes 'The study of the specific effects of the geographical environment, consciously organised or not, on the emotions and behaviour of individuals'.

(Merlin Coverley. *Psychogeography*. Hertfordshire: Pocket Essentials, Harpenden, 2006, p. 10)

18 There is a belief that European expansionism via colonialism throughout the world brought a great deal of benefits to countries under their control such as infrastructure and centralized governance in the continents, regions, and islands of North, Central and South America, Africa, Asia, Australia, and Oceania. But as is clear these benefits were for the sole purpose of effectively administering resource extraction, cultural destruction, enslavement, forced labour, forced abandonment of spiritual beliefs, displacement, and subjugation to repressive racial laws. It cannot be ignored that colonialism and industrialization are directly linked to European wealth in the 18th and 19th centuries and remained so in the 20th and into the 21st centuries. It also cannot be ignored that colonized countries have not received appropriate reparations from the excesses of their former colonial invaders and the wealth accrued from the theft of resources.
19 See Frantz Fanon. *The Wretched of the Earth*, translated by Richard Philcox. New York: Grove Press, 1963, p. 6.
20 Ibid., pp. 54 and 56.
21 See Jean-Paul Sartre's preface to Fanon, *The Wretched of the Earth*, p. xliii.
22 English carpenter and cloth-maker John Harrison (1693–1776) spent 30 years perfecting the chronometer to solve the problem of determining longitude, which was at the time the largest advancement in navigating the world's oceans. The chronometer, a mechanical clock designed to pair time and distance travelled turned getting lost in the vastness of the ocean into the realm of locational spaces of the map and as a result to land and continents. See Dava Sobel. *Longitude: The*

Story of a Lone Genius Who Solved the Greatest of His Time. London: Fourth Estate, 1996.

23 See Marie Battiste, ed. 'The Context of the State of Nature'. In *Reclaiming Indigenous Voice and Vision*. Vancouver: UBC Press, 2000, p. 31.

24 Ibid., p. 71.

25 A case in point concerning the immense suffering of African people at the hands of their colonial rulers is the atrocious 23-year rule by Belgium of the then called Belgium Congo (now DRC) in the latter half of the 19th century. Ten million Congolese are said to have either been massacred or died from disease and famine. The other prime example was the apartheid system in South Africa that only ended in 1994. Millions of South Africans were forced from their lands and moved to townships and suffered extreme discrimination from white Afrikaners. Since the end of apartheid in 1994, the remnants of racial segregation and oppression still haunt many South Africans, their lives and self-determination, and access to their former tribal lands.

26 See Nnimmo Bassey. *To Cook a Continent: Destructive Extraction and the Climate Crisis in Africa*. Cape Town: Pambazuka Press, 2012, p. 5.

27 Ibid., p. 5.

28 See David Morley. *Home Territories: Media, Mobility and Identity*. London: Routledge, 2000, p. 182.

29 Ibid., p. 183.

30 See Stephen Castle and Mark J. Miller. *The Age of Migration: International Population Movements in the Modern World*. London: Macmillan, 1998, p. 104.

31 See Willem van Schendel and Itty Abraham, eds. *Illicit Flows and Criminal Things: States, Borders, and the Other Side of Globalization*. Bloomington: Indiana University Press, 2005, p. 41.

32 See Jennifer Gabrys. *Digital Rubbish: A Natural History of Electronics*. Ann Arbor: University of Michigan Press, 2013, p. 58.

Bibliography

Bassey, Nnimmo. *To Cook a Continent: Destructive Extraction and the Climate Crisis in Africa*. Cape Town: Pambazuka Press, 2012.

Battiste, Marie, ed. 'The Context of the State of Nature'. In *Reclaiming Indigenous Voice and Vision*. Vancouver: UBC Press, 2000, pp. 11–38.

Bourke, Joanna. *Fear: A Cultural History*. Emeryville, CA: Shoemaker and Hoard, 2006.

Castle, Stephen, and Miller, Mark J. *The Age of Migration: International Population Movements in the Modern World*. London: Macmillan, 1998.

Coverley, Merlin. *Psychogeography*. Hertfordshire: Pocket Essentials, Harpenden, 2006.

Cowen, Leah. *Border Nation: A Story of Migration*. London: Pluto Press, 2021.

Di Cesare, Donatella. *Resident Foreigners: A Philosophy of Migration*. Translated by David Broder. Cambridge: Polity Press, 2020.

Fanon, Frantz. *The Wretched of the Earth*. Translated by Richard Philcox. New York: Grove Press, 1963.

Gabrys, Jennifer. *Digital Rubbish: A Natural History of Electronics*. Ann Arbor: University of Michigan Press, 2013.

Lucassen, Jan, Lucassen, Leo, and Manning, Patrick, eds. *Migration History in World History*. Boston: Brill, 2010.

Morley, David. *Home Territories: Media, Mobility and Identity*. London: Routledge, 2000.

Mountz, Alison. 'Refugees—Performing Distinction: Paradoxical Positionings of the Displaced'. In *Geographies of Mobilities: Practices, Spaces, Subjects*, edited by Tim Cresswell and Peter Meriman. Farnham: Ashgate, 2011, pp. 255–269.

Schendel, Willem van, and Abraham, Itty, eds. *Illicit Flows and Criminal Things: States, Borders, and the Other Side of Globalization*. Bloomington: Indiana University Press, 2005.

van der Pijl, Kees. *Nomads, Empires, States*. London: Pluto Press, 2007.

3 Earth Extractions
Pillage and Ransack

Plunder—Exploitation and Destruction

'The climate timebomb is ticking'.

'Greenhouse gas emissions keep growing. Global temperatures keep rising. And our planet is fast approaching tipping points that will make climate chaos irreversible. We are on a highway to climate hell with our foot on the accelerator'.

'We have a choice. Collective action or collective suicide. It is in our hands'.

'As a global community, we face a choice. Do we want migration to be a source of prosperity and international solidarity, or a byword for inhumanity and social friction?'
(Selected quotes from various speeches delivered by United Nations Secretary General Antonio Guterres between 2022 and 2023)

Throughout his tenure as United Nations Secretary General, Antonio Guterres has ratcheted up his rhetoric in respect of the most pressing issue facing humankind: its continuing survival. Breaking from the usual code of soft diplomacy, Guterres' outspoken criticism has jolted the international community into acting now so that future generations will have a livable Earth to inhabit. While many listen, the powerful forces of fossil fuel companies, climate change deniers, and some governments remain deaf to his pleas. Climate change has become a battle of words played out between countries, governments, citizens, corporations, and their shareholders, and in the middle sits the one word intended to save us all: *sustainability*.

In his introduction to *Comparative Planetology*, Lukáš Likavčan cites the US Department of Energy's rebranding of natural gas in March 2019 as 'molecules of freedom'. The spin employed by the Department of Energy is designed to appeal to our susceptibility.

Likavčan suggests to understanding 'how the chemistry of our planet evaporates our old modes of political thinking'. The idea is to believe that the ongoing extraction and use of natural gas facilitates energy 'freedom' on account of its molecular invisibility. Referred to as 'greenwashing', this

DOI: 10.4324/9781003382515-4

palatable rebranding reveals the ease with which government institutions and companies can rebrand Earth-damaging substances as environmentally friendly despite the impact they cause. 'Climate emergency shows us that if chemistry is political, politics is also chemical; or in other words, politics always involves the operation and manipulation of chemical compounds and processes'. Likavčan equates the politics of energy 'to some set of chemical procedures' where

> politics as we know it is contested by the fluid, dynamic and precarious realities of politics-to-come, where every action can be read as a chemical process in the planetary ecosystem, since it is linked—directly or indirectly—to carbon emissions, metabolism of methane and nitrogen, acidification of the oceans, and so on.[1]

This meshing of politics and chemistry with the extraction and burning of fossil fuels reveals how the power of marketing can undermine attempts to reduce CO_2 emissions and reach net-zero targets.

Doctoring the truth and shirking responsibility is evident across the fossil fuel industry. For example, the oil industry's widespread use of gas flaring as a safety measure to avert the risk of an explosion involves burning off natural gas to reduce gas pressure in underground oil reservoirs, but this is not the only reason behind the practice. Oil companies burn off natural gas to avoid the cost of collecting, transporting, and storing, which they claim is economically unviable. There are significant health hazards attached to gas flaring for anyone living near oil fields. The Nahran Omar oil field near the city of Basra in southern Iraq has one of the highest concentrations of toxic chemicals as a result of gas flaring in the world.[2] One of the toxic chemicals is benzene, which is known to heighten the possibility of cancer, particularly leukaemia. A joint venture between the Basra Oil Company, BP, PetroChina, and SOMO has disputed any connection and responsibility of gas flaring on the nearby population. The health risks on humans would be obvious in the continuous pouring of black toxic smoke from chimneys of burning gas. In 2021, it was estimated that gas flaring significantly contributed to climate change; releasing 400 million tons of carbon dioxide into the atmosphere.[3] To put this figure into perspective, it amounts to more than the United Kingdom's total carbon emissions for that year. Gas flaring also releases other hazardous chemicals such as naphthalene and black carbon, both of which cause considerable breathing problems and serious growth issues in young children. In *Under Poisoned Skies*, documentary filmmaker Jess Kelly exposes the effects that gas flaring is having on residents living near to the Nahran Omar oil field. Asked how she become aware of the health effects of gas flaring on the community, Kelly replied: 'we first became interested in the toxic effects of gas flaring after seeing shocking videos on social media of gas flaring in southern Iraq and reports from communities living near oil fields like Nahran Omar, where BP is the lead contractor, that cancer was so common it was like the flu'.[4] To bypass responsibility and avoid any possible legal action, the companies who own the Nahran Omar oil field are not registered as managing its day-to-day operations and through this loophole relieve their accountability with regard to the connection between gas flaring and health risks posed to the surrounding population. Gas flaring is just one example of how fossil fuel corporations exchange their 'molecules of freedom' for huge profits in contrast to the wellbeing of human life and life of the planet.

Avoiding their responsibility for the health risks associated with gas flaring illustrates how fossil fuel companies are able to systematically abuse international human rights laws that are intended to protect human life, showing that such rights can be forfeited if there are big enough profits to be made. Global action in the face of climate change across the world is poised between a powerful minority able to manipulate international laws, agreements, and responsibilities in the pursuit of the highest profit margins possible, and the majority of people who suffer the environmental and economic devastation and health issues as a result. With huge financial resources at their disposal and political clout, fossil fuel corporations maintain their support within government circles to withstand demands for responsible climate change

Figure 3.1 Cover image, gas flaring, Nahran Omar oil field, Iraq.

Source: Image still from documentary film *Under Poisoned Skies*, director Jess Kelly, producer Owen Pinnell for BBC Arabic Investigates, 2022.

action. The climate crisis is divided: on the one side, there is the maintenance of capital growth and wealth; on the other, there is environmental destruction and poverty. Compounding the problem are the divisive tactics of the right and left of the political establishment of individual countries, the wider geopolitical world order and well below both of these, people and the health of the planet. This polarized world assists fossil fuel companies for it affects 'the prospects of planetary cooperation in the wake of climate emergency', Likavčan writes, 'since they complicate political geography by preferring fragmentation over integration'.[5]

As cited in the previous chapter, 'Terrestrial Migrations', global extraction of the earth's resources was industrialized and perfected through the forces of dispossession, suppression, and enslavement under colonialism. The resulting upheaval, loss of independence, self-determination, homelands, and the shattering of traditional cultures also ushered in a new phenomenon of global inequality. Global inequality would become industrialized in the modern era as the human element in undertaking the work of resource extraction was removed and replaced with the tireless worker of mechanized machinery. In *Planetary Mine*, Martin Arboleda explains how the modernization of mineral extraction via mechanization brought about an unlimited capacity for environmental destruction.

> Although in the popular imagination mining is usually considered to be a rudimentary activity, the degree of technological sophistication that mediates the extraction of minerals from the subsoil in the twenty-first century is nothing short of astonishing. Innovations in artificial intelligence, big data, and robotics have allowed mining companies to introduce automatic trucks, drills, shovels, and locomotives to the stages of the production process. Some of these sophisticated machineries—most notably trucks and shovels—are not remotely controlled; they are fully robotized, which means that they can operate twenty-four hours a day, seven days a week, without direct human intervention.[6]

Removing the human element from the industrial extraction of minerals to artificially intelligent machines also removes the responsibility for environmental devastation. This avoidance can be similarly understood for example in the use of drone strikes on enemy targets, which remove the soldier from the fatalities incurred by their location thousands of kilometres from the field of battle. Terrestrial extraction of the earth's resources became operationally virtual as drone strikes in war became violently virtual. Topographies and geographies of regions and continents have been altered by resource mining, transforming terrain into landscapes of desolate extracted surfaces. Arboleda writes that, in the mining industry,

> the self-objectifying practice of capital has metamorphosed into rivers poisoned by mercury and cyanide, ancient glaciers torn to shreds by

explosives and machineries of extraction, peasants ravaged by debt, police forces bizarrely out of control, mining towns riddled by cancer epidemics, and rampant labor casualization.

Mining extends far beyond just mineral extraction and environmental destruction of landscapes; it includes the plunder and extraction of bodies, 'rendering visible how human bodies become possessed (and often obliterated) by uncanny forces and nonhuman objects become animated with powers over life and death'.[7] Where industrialized mining tears apart the environment and displaces the original owners of the land, for others it brings benefits in resources being converted into products consumed in countries far from the sites of extraction and environmental devastation. Scarred ground, environmental obliteration, and product rewards and profits keep the wheels of industrial manufacturing turning. Widely established during colonialization and perfected in the 19th and 20th centuries, the transnational trade later expanded through globalization has brought little benefit to those people situated where a great deal of resource extraction is undertaken.

Human disruption of the earth did not begin with the economic and racial polarization under colonialism nor the imbalances of globalization between developed and developing countries, it began with humanity's historical connections to survival. Modern humans live on the histories of early human relations with the earth founded over hundreds of thousands of years of roaming, hunting, and gathering to a cultivated relation in the establishment of settlement 12,000 years ago. As explored in depth in Chapter 2, settlement recalibrated the earth to a geographical human worldview and its resources to geological processing. Over successive millennia, human relations with the shaping of geography progressed and then exponentially sped up in the last 200 years in the industrial use of coal for the smelting of minerals, oil and gas extraction to supply energy production, industry and modern-day transport. In 200 years, humanity has reshaped its historical connections to the natural environment from manual labour to mechanical industrial proficiency and in doing so it has begun a process of disaffection and disassociation from nature. From settlement to empire, the Age of Exploration to colonialism, industrialization, and modern corporate imperialism, resource extraction has shaped environments and sharpened the global geographical, geological, and economic divisions between peoples and continents we see today. The environmental consequences of this division—wealth inequity; land degradation; deforestation; drought; air, sea, and river pollution—continue as a result of the geographical separation between suppliers and consumers. Where the establishment of agriculture shaped terrain to guarantee food supplies for human survival, resource extraction went further by altering topographies, whole regions, and populations in the name of human progress.

The plundering of the earth's resources brought an economic currency in terms of valuing a region's resources in conjunction with the expendability of the surrounding environment. Scanning the earth's mineral-rich areas such

Figure 3.2 Open-cut mine.

Source: Image courtesy of Vlad Chețan, Pexels.

as Africa, Australia, Brazil, Russia, and the United States reveals mining operations that have flattened topography and gouged the terrain into spirals burrowing deep into the earth. Hydraulic shovels capable of excavating 50 tons of minerals in one go, driverless trucks conveying coal and ore, pipes and ships transporting oil and gas and steel blast furnaces, chemical and manufacturing companies burning and transforming mineral elements into energy, construction materials, and products—all of this feeds the industries and lives of people in the modern world.[8] A satellite survey of the world's surface reveals humanity's physical impact on the earth. Images of decimated landscapes due to bauxite, iron ore, zinc, tin, nickel, and gold by multinational corporations such as Vale, Anglo American, and Rio Tinto in the Brazilian Amazon have been well known for decades, yet their destruction of the environment to extract resources continues unabated.[9] In the Congo Basin in the DRC, bauxite, diamond, copper, uranium, and gold mining are rapidly erasing large tracks of the world's second-largest rainforest.[10] In the state of Borneo in Indonesia and in Papua New Guinea, the natural

Figure 3.3 Articulating shovel coal mining machinery.

Source: Image courtesy of Pixabay, Pexels.

rainforest is being felled and, in its place, huge plantations of palm oil are being installed, resulting in severe ecological, animal, plant, soil, and water degradation.[11] All three of these major natural ecological environments share an apocalyptic image of a scarred Earth to accompany the desertification of

Figure 3.4 Drone image, palm oil plantation, Merauke, Papua New Guinea, 2015.

Source: Image courtesy of Sophie Chao.

landscapes, dried-up rivers and lakes, melting glaciers, and sea ice regions throughout the world. Humanity's gross ecological violations on the earth's surface illustrates its ability to shape nature to satisfy its inexhaustible consumerism, even at the risk of bringing about its own extinction.

In the 2020 report of the United Nations Convention to Combat Desertification (UNCCD), it is stated that 70% of landmass had been altered and up to 40% degraded due to human impact such as the expansion of agriculture. Surprisingly, the report is only the second undertaken by the United Nations and in response to its findings, the stated objective is to reduce land and ecosystems degradation by coordinating an international response to remediate 1 billion hectares (10 million km²) of land by 2030.

> Land and ecosystem restoration will help slow global warming, reduce the risk, scale, frequency, and intensity of disasters (e.g., pandemics, drought, floods), and facilitate the recovery of critical biodiversity habitat and ecological connectivity to avoid extinctions and restore the unimpeded movement of species and the flow of natural processes that sustain life on Earth.[12]

For the UNCCD to reach this target, it will have to contend with a number of issues driving mass land clearance such as human population expansion, intense farming, and various government subsides, policies, and lack of regulation of the mining, fossil fuel, and logging industries. By 2050, the world's population is predicted to hit 10.4 billion people—a rise of 2.4 billion on the current 8 billion reached at the beginning of 2023. Ongoing land clearances, such as in Borneo, Indonesia, Papua New Guinea, the Congo Basin, Mexico, and the Amazon Rainforest stretching across Brazil, Colombia, Peru, and Bolivia amounting to millions of hectares, place the UNCCD's land restoration target in a seesaw of destruction and restitution. As of 2018, annual global land degradation amounted to approximately 12 million hectares or 120,000 km². Doing the maths, the 8-year UNCCD timetable to restore 1 billion hectares (10 million km²) would appear somewhat fanciful, not to mention the difficulties in organizing international cooperation and securing the funds required. Not to dismiss the UNCCD's determination to achieve its goal, but its real fight is an economic system that places the exploitation of resources for profit over the health of the planet and human life.

Humanity's perverse propensity for destruction of the natural environment can also be seen in its relations with wild animals such as elephants, lions, tigers, rhinos, and many once prolific and now endangered species. At the turn of the 19th century the global elephant population sat around 10 million. In 2022 the estimated global elephant population stood at 497,000 and out of this population 400,000 are African elephants where approximately 30,000 are killed every year. This catastrophic figure is disturbing but when considered alongside the average extinction of animal and plant species worldwide, which is estimated between 150 and 200 a day, it fits into humanity's propensity for annihilation whether it be humans, animals, plant life, or the earth. The UNCCD's land restoration target might be reassuring in countering land degradation, but what is less reassuring is humanity's capacity to realize such a goal given its capability to bring about its own destruction.

As cited in Chapter 2, nomadic life relied on an embedded ecology in connection with the surroundings. With the establishment of settlement, the switch from terrestrial nomad to sedate urban dweller, humanity's ecological connection to the natural environment was progressively expelled from the body. 'The figure of the Terrestrial stands in opposition to the Globe', Likavčan writes, through

> the combination of various visual, philosophical, and (geo)political revolts against globalist ways of imagining Earth. [T]he Planetary is juxtaposed with both the Globe and the Terrestrial, overcoming theoretical, political, and ecological obstacles emerging from these conceptualisations, and instead treating the planet as a geophysical, impersonal process.[13]

The traction of the 'Terrestrial in opposition to the Globe', for Likavčan 'might render anew our present approach to climate change'.

One of these visions is the figure of Earth-without-us, which serves as a normative template for political and design interventions that might become the vectors of an upcoming period of post-Anthropocene. The second figure based on the Planetary that comparative planetology formulates is Spectral Earth, addressing our relation to extinction and species-being. Both Earth-without-us and Spectral Earth allow us to address the geopolitical dimension of climate emergency in relation to our visual and philosophical imagination of the planet without referring to figures drawn from colonial histories or reactionary, identitarian tendencies in contemporary geopolitics.[14]

Geopolitics was not in the minds of early humans' comprehension of the world and even if it were, it would have been an earthly, geo-local connection. The planetary actions of fossil fuel, mining companies, illegal loggers, cattle and plantations owners to deforest, mine, and extract over huge tracks of land, even if inadmissible, has become part of the 'unacceptable' destruction of the earth in which humans are engaged.

Humanity's capacity for conception and innovation set it apart from all other living things as a natural source of its progression. From the discovery of fire and invention of tools from the plough to the front-end loader, humans have shaped the earth perhaps more than they have been shaped by it. In *Flatness*, W. B. Higman explains that humans shaped their environments by ruffling out the contours to smooth flat planes 'on which human beings can comfortably walk, drive, communicate and play; in the pursuit of profit and pleasure; and in the making of civilization' that characterize 'everyday life on earth'.[15] The technologies humans deployed to advance their progression were also used for destruction—as is the case now as it was in ancient times. In book 10 of *De Architectura*, the Roman architect and military engineer Marcus Vitruvius (1st century BC) refers to agricultural inventions in the same vein as military contraptions where hoisting machines, water wheels, and windmills are equal to catapults and siege machines, each assisting in cultivating land or fighting wars. Taken to their fullest potential, machines for agriculture altered the natural environment in the same way that machines for war threatened human beings in their ability to destroy one another. Both were programmed to permanently harness and alter the trajectory of human progression. Vitruvius saw no separation concerning machine invention; where every field ploughed and citadel conquered, the machine provided humans with the benefits of productivity.[16]

When animals were deployed as the early machines for land cultivation, their secession to mechanical devices gave humans the ability to vastly expand their dominance over the natural world. Before such a time, humans and animals had combined their efforts to turn rugged terrain into fields of productivity through a shared exhaustive manual labour. In the timeline of human life on the earth to the invention of the plough approximately

4,000 years ago in ancient Egypt and Sumar to the driver (if still needed) at the wheels of excavation machines in the 19th, 20th, and 21st centuries, machine and human dominance of the earth are now inseparable. The once invincible scale of geography and terrain to the ox and its handler can be conquered and tailored to the needs of humanity by elevating the machine as our near equals. Machines became an extension of humans in the tasks of shaping and gouging ground to extract minerals, ploughing for crops, flattening topographies to build cities, rerouting rivers, and damming lakes. With the ease of pressing buttons and operating levers, machines aided modern human life to the point that it now cannot exist without them, which is also to say that human life cannot exist without the destruction of the earth.

In *Humans, Animals, Machines: Blurring Boundaries*, Glenn A. Mazis refers to machines as the 'postmodern animal'. 'In the twenty-first century', humanity has been haunted by 'the images of our souls, as an unknown face of who we are and who we are becoming...a repressed dimension harboring fears...just as the animal haunted our forebear's dreams, fears, and desires, and still does in ways unnoticed by most of us'.[17] Humans are no longer haunted by animals or of themselves; instead they are now haunted by the earth and by what they have made—excessive exploitation that has wrought global warming and climate change upon the planet. Ecological sustainability was inherent in nomadic life but through settlement the earth became a supplier for everything imaginable via the cockpit of a machine. The wounds humans have inflicted on the earth from deforestation to spiralling excavation of open-pit mining is not the fault of the machine or the operator, but rather the idea of radically shaping human life according to unlimited desires. Humans now encircle the earth with their machines, impacting ground and threatening ecologies to alter the course of environmental evolution for a more powerful evolution of the techno-human.

The machine age intensified the plundering of resources to fire the furnaces, forge the presses, and process minerals in the manufacturing plants across the world. The rapid transformation to industrialization turned the planet into a global resources market; towns and cities sprung up where minerals were mined and forests cleared for their timber. Industrialization required industrial-scale resource extraction and resource-rich but economically poor countries provided a cheap supply of labour for the taking. Returning to Bassey's *To Cook a Continent*, cited in the previous chapter, he refers to the plundering of the African continent as an 'irresistible' enterprise.

Access to raw materials and cheap labour made the plunder of Africa irresistible. One slave owner was quoted as saying that slaves were 'free'; all you needed was to gather them in. Bloodletting and easy dispensation of native lives meant nothing. Thus, the early drive into Africa was fueled by a liberty to do as one pleased within the sandwich of commerce and conquest.[18]

This irresistible double 'gift'—free resources and free labour for the taking—became an acceptable part of industrialization, European expansionism, and wealth accumulation. If African lives 'meant nothing', so too did the environment. To squeeze every drop of the continent's resources with every drop of blood of slave labour not only set in motion the disregard of human life and the environment but also enshrined inequality as a standard practice for human progress. As noted in the Introduction and Chapter 2, the climate crisis of today has its roots in the bloodletting of colonial invasion and early industrialization. While accepting that much has changed since then, specifically freedom and independence from colonial rule, very little has changed in terms of global equality, which is translated today into the terms of global sustainability and is by and large a selective practice, where the poorest but resource-rich nations are excluded.

The vast global 'colonial' backyard set up in the 18th, 19th, and 20th centuries by rich nations, which wanted to acquire resource-rich regions by force, was simply a repository to generate wealth and use 'captive' human labour to extract it. Environmental activist Vandana Shiva writes: 'Development' becomes a strategy to 'combat scarcity and dominate nature to generate material abundance. Capital accumulation through appropriation of nature is seen...as a source of generating material abundance, and through it, conditions of peace'.[19] Peace was not given in occupied lands during the colonial era, just as it is not today where mercenary armies control regions, mines, and labour on foreign lands. In *The Robbery of Nature*, John Bellamy Foster and Brett Clark observe how the expropriation of nature,

> since human beings are inherently a part of nature, is undermining the natural–material bases on which humanity's existence rests. This degradation of the human relation to the earth results from treating 'Nature' as a 'free gift to...capital,' and from the violation of the basic 'conditions of reproduction' and ecological sustainability.

Bellamy Foster and Clark continue this course of reasoning, saying that '[t]he expropriation of nature is at one and the same time the expropriation of land/ecology and the expropriation of human bodies themselves'.[20] The authors argue that human expropriation of land and ecology is problematic in 21st-century capitalism for the reason of the economic and societal revolution required to reverse it.

> The means for the creation of a just and sustainable world currently exist, and are to be found lying hidden in the growing gap between what could be achieved with the resources already available to us, and what the prevailing social order allows us to accomplish. It is this latent potential for a quite different human metabolism with nature that offers the master key to a workable ecological 'exit strategy'.[21]

If global economics continues to be founded on the plundering of the earth's resources, then sustainability as the ecological goal to secure the future of human survival will fall out of reach. The logic of capitalism demands unfettered plundering of the earth and global economic inequity to drive it. This system was established over centuries of European exploration and colonial and Western economic expansion. Overcoming humanity's present estrangement from nature and to balance the resource plundering of the earth in order to maintain economic growth puts sustainability in a fragile position in terms of its capacity to counter the earth's environmental decline. Finding an ecological 'exit strategy' to the effects of climate change will continue to flounder when the colonial powers of the past live on through the forces of corporate imperialism today.

Estrangement—Antagonism, Disaffection, Hostility

First by the hand with pick and shovel to automative machines of immense capacity, the extraction of the earth's resources catapulted an estrangement between humanity and the natural environment. It gave rise to one or more reactions such as antagonism, disaffection, hostility, isolation, and alienation, prompting an inability to be part of something or connect to one's surroundings. The origins of human estrangement do not follow a certain course. For early humans, estrangement may have been a natural cause of survival in dealing with the threats posed by the natural environment. Living in accordance with nature, early humans were at its mercy until the formation of settlement began nature's exclusion as the dominant factor determining life and death. The construction of settlement forged new relationships to the natural world, shifting human interaction from the familiar to the unknown outside its walls. Imagined fears and threats became a by-product of sedentary occupation, which continues today within as much as outside cities to the territorial borders of countries.

Estrangement can rise at any moment. Walking through a forest can spur a sense of estrangement where dappled light and darkness diffuse the trail. Without knowing it, the walker has wandered off the trail and confusion sets in as to which direction to take. Soon an overpowering sense of helplessness and imagined fears and threats creep in and the forest now turns against them becoming sameness, seamless, and limitless. Shadows now play upon the wanderer's growing fear and survival now occupies their mind to apply untested skills to conjure their memory back to the track they accidentally departed. As night falls, the wanderer is forced to accept their failure and beds down under a tree, exhausted, frightened, and cold. As the eerie sounds of the forest increase in intensity, the wanderer lays awake to the morning light and a new opportunity to walk their way out. As night falls on the second day, the wanderer has unknowingly been walking in a circle and has returned to the same spot where they spent the previous night. While the familiarity of the tree brings a sense of relief, it also increases their sense of estrangement. With another night spent lying awake, the estrangement can only be broken by daybreak bringing a welcomed relief and familiarity of their helplessness.

It is clear in this story that the lost wanderer's estrangement from the forest, which started when they first became, is now determined by their imagination. The walker is now foreign to the forest and the forest alien to them. The problem the wanderer faces is how to transfer what they know outside the forest to inside it. This may have been what early nomads experienced in their migratory journeys across the earth. Nomadic life was constant adaptation of the unknown and transformation to their surroundings. Through innovation and design, humans sought ways to struggle less with the natural world as a just cause for their advancement. Along the epic narrative of human progression came the ability to transform their surroundings rather than be transformed by them and this provided unlimited possibilities. Aided by the tools of mechanization, humanity wrestled with the earth for dominance, and as it became more successful in doing so, its estrangement from nature grew. Estrangement had moved from threat and fear of the natural world to disaffection and separation for control. With the immense benefits that accompanied the industrialization and manufacturing of the 19th and 20th centuries, humans were now foreseeing the emergence of their own demise.

Environmental estrangement from the natural world is felt and experienced differently between the world's urban and rural populations. The transition of rural populations into urban centres increased human defamiliarization with the natural environment, precipitating the psychological estrangement that exists in urban centres today. This exodus of rural populations into the spaces of human made control desensitized human connectivity with the natural environment. As cities are the supreme representation of humanity's technological achievement and societal construction wrapped in phantasmagoric desire, entertainment, and wonderment, humans had found they could equal what existed in the natural world. Cities herald humanity's ascendency over all living things. In *Readings of Exile and Estrangement*, Julia Kristeva and Anna Smith note that 'the means we have for registering information about the world that surrounds us is constrained and filtered through the screen of habit'.[22] Constructing habitual behaviour is one of the hallmarks of cities, first physically in removing topography for a rigid layout to manage capital flow, labour, and living. Cities were designed to ritualize the habits of the urban dweller between work, home, and entertainment. To do this, city planners removed any sign of the natural environment, controlled sanitation, and water supplies and denatured essential items such as food production in favour of their import on supermarket shelves. In short, the value of the city was determined by the extent of devaluing the natural environment. The progression from nomadic life to settlement to the city programmed nature from cooperative exchange to estrangement. A veneer of constructed surfaces designed to resist storms, rain, wind, heat, and cold further increased the human capacity for estrangement from the natural world. In rural regions the experience of the natural world was the opposite; theirs was one of dependency on exposure to maintain their survival. The impact of climate change has now changed this dependency and destabilized human survival. Recurring catastrophic weather events such as drought, floods, soaring heat leading to water scarcity, desertification, and acidification forged an unnatural estrangement between humankind and the natural world. No longer can rural populations rely on a habitual nature in which to securely grow their crops as urban dwellers rely on the supermarkets. The veneer of the city designed to estrange nature from the daily experience of its occupants has expanded to the rural regions under climate change.

Not addressing the disparity in CO_2 emissions between urban and rural populations, the core environmental ideology of sustainability cannot succeed when the dislocation between these populations continues to widen. If we were to believe that through climate change environmental estrangement has become equalized between rural and urban populations, it still remains disproportionate to the causes. To illustrate, the African continent could be considered the most visible example of the effects of global warming. Home to 1.4 billion people, the African continent is responsible for just 3.7% of global CO_2 emissions, which amounted in 2021 to 436 million metric tons of CO_2. In 2021, global CO_2 emissions amounted to 38 billion metric tons.[23] To understand this statistic in environmental terms in relation to how the earth absorbs CO_2 emissions, the Amazon rainforest processes through photosynthesis 4% or 1.5 billion tons of CO_2 emissions a year.[24] To put this fact into the context of people, of the world's population of 8 billion people, 3.6 billion live in regions affected by climate change. It is clear that to tackle the effects of climate change across the world, the gap between rural and urban dwellers must be reduced. Present global sustainable policies hinge on international agreements to decrease global warming to eliminate fossil fuel dependency. Setting zero carbon target emissions, increasing renewable energy sources (solar, wind, hydrogen), eradicating plastic packaging, reducing environmental damage in the extraction of mineral resources, initiating land-restoration projects, moving to clean industry manufacturing, and reducing consumer consumption levels are both urban and rural based. To achieve these ends, the global financial system has to be reset and moved away from the continuous economic growth of cities and consumption to the less profitable but integral benefit and support of rural populations.

Estrangement—antagonism, disaffection, and hostility to the natural world—set the course of human advancement. European colonialization of the Americas, Africa, Australia, and parts of Asia set antagonism, disaffection, and hostility as an integral part of invasion and the plundering of resources of foreign lands and people. What humans fail to see then and only recently now understand is the interconnectedness of the earth's ecological regions. One of the early exponents of the earth's interconnectedness was

UK environmentalist, James Lovelock, who formulated his Gaia theory 'in which all life and all the material parts of the Earth's surface make up a single system, a kind of mega-organism, and a living planet'.[25] Explaining the earth as a unified system of which humans are but a part, Lovelock's Gaia theory is central to maintaining a balance—the biosphere of all living and non-living things. 'The biosphere is the three-dimensional geographic region where living organisms exist', Lovelock explains, and 'Gaia is the superorganism composed of all life tightly coupled with the air, the oceans, and the surface rocks'.[26] Lovelock's theory of a superorganism *Gaia* of the earth, where the impact on one element affects all connecting elements, was, for a substantial period of his advocacy, dismissed by environmental scientists as empirically unscientific and by governments as a thorn to economic growth. Lovelock presented a way of thinking for humanity to redress its estrangement from the natural world and to reposition itself as a part of the superorganism of the earth through a sustainable partnership. Lovelock's advocacy to shift human perception to an integral 'family' with the earth challenges the system of disaffection humanity has built up until now. To achieve humankind's link to the earth's Gaia is yet to materialize and as time tips the balance further away, Earth estrangement and habitat destruction will continue to increase.

Overturning the devastation of human impact on the earth to the restoration of the earth's Gaia is ultimately the measure of humanity's capacity for the 'care of the self' as the care of the oceans, continents, forests, glaciers, animals, plants, and the atmosphere. For this to seep into the consciousness of human awareness and create the capacity for a radical alteration in human relations with the earth, it is not enough to rely on technological innovation to combat the effects of climate change all the while sticking to the same path in mass extraction and processing of the earth's resources. For humanity to be absorbed within the Gaia of the earth is dependent on detangling from its habitual estrangement consistent with humanity's advancement and devastation of vast tracks of the earth's surface. Environmental devastation has allowed humanity to achieve immense benefits of wealth and comfort at the same time as it has sustained global poverty and hardship. Redistributing global wealth would no doubt speed climate change recovery for it would spread responsible capability for adaptation and building resistant infrastructures. To do this would mean re-evaluating human progress from an estranged relationship to the natural world determined by resources extraction and consumption to one inherently connected. The pathway to sustainable practices that has so far been pursued is built on the belief that in order to 'fix' climate change and restore Earth health, this means confronting the earth's devastation from the distance of human innovation and technology. The pathway to achieve global sustainability is not only to disable human estrangement from the natural world but also to establish a global population capable of adapting, transforming, and connecting urban and rural dwellers and, if human imagination allows, to be absorbed into the earth to become a single entity within the superorganism of the earth as Lovelock advocated.

The weakness of sustainability as a global call to live sustainable lives is to keep doing what we are doing but to do it differently—in other words, to do the same thing but in a more environmentally friendly way. But this does not deal with the origins of humans' estrangement in relation to the natural world. The effects of human impact on Earth will be felt for decades to come, but there is a possibility to avert human dominance and instil the capability for human adaptation and transformation. As will be further discussed in Chapter 4, global sustainability is possible if humans claim their own ecology—plant, animal, and human life combined—that over time reconnects to the evolutionary movement of transmutation and adaptation of all life on the earth. If humanity is able to reach this point, the destructive elements of global warming and climate change might become exhausted, atmospheric turmoil subdued, and weather regulated to predictable patterns. Future life on the earth depends on the capacity of humans to reimagine their existence and at the centre of this is the removal of our estrangement from the natural world. A new set of tools must be designed and new human life skills implemented to globally combine people and the earth into a single transforming entity. To do this, much has to be given up—that is, all the things that presently divide humans and humans from the earth.

Equality—The Matter of Sustainability

In his book, *A Brief History of Equality*, Thomas Piketty argues that societies have been progressively moving towards greater equality. He points to the devolution of powers responsible for systemic inequality such as monarchial power and colonialism to the rise of democracy, socialism, health, and education. This being true, it nevertheless has not been transferred to an equitable response in confronting the climate crisis. 'Without resolute action seeking to drastically compress socioeconomic inequalities, there is no solution to the environmental and climate crisis', he informs us.

> To make progress in this direction, we must combine different indicators—environmental and economic, for example—and independently set targets for carbon emissions or biodiversity while at the same time formulating objectives that include the reduction of inequalities in income and the distribution of fiscal and social deductions and public expenditures. In this way we may compare different sets of public policies that make it possible to achieve our environmental objectives.[27]

Inequality has been an accepted flaw in human history in terms of one civilization or race dominating and suppressing another. Over millennia, people who have been subjugated to inequality have staged insurgences such as the slave rebellions during the Roman Empire and the peasant revolts that raged across Europe during the medieval period. Many of these revolts were relatively short-lived due to them being crushed by the far superior military might of the rulers. The 18th and 19th centuries were characterized by the broadening of inequality across the world in the form of racial inequality managed through the system of colonialism, as previously discussed. This would eventually be broken in the 20th century in Africa, South America, South-East Asia, and Oceania through countless bloody wars of independence, but the inequality would remain. In the 18th- and 19th-century European and English societies, the rise of industrialization saw great class upheavals. While the merchant class wrestled wealth and commercial power away from the ruling classes, the working class, whose labour powered the factories, was placed under institutional and policing controls.

Inequality was established through the dispossession of land via the swords and guns of invaders, and when the swords and guns fell into the hands of the enslaved and oppressed, they fought for their freedom. Racial inequality was an accepted consequence under colonialism, just as the corporate imperialism of transnational companies is an accepted result of globalization today. A great deal of racial inequality was a controlled system devised by white people over black, brown, and Asian peoples with the aim of acquiring resources and cheap labour by any means necessary. For example, the slave trade in which millions of Africans were rounded up and shipped to work on cotton and sugar plantations in Brazil, the Caribbean, and America was maintained by a constant supply of slaves to replace deceased slaves. Plantation owners viewed their slaves as less than people and even less than animals. The shameful legacies of racial subjugation, slavery, and death are still with us in the more closeted geopolitics and economics of globalization. Piketty acknowledges that the history of colonialism has not passed, acknowledging that

> it would be naïve to imagine that its effects can be erased in a few decades. Those born today are not individually responsible for this burdensome heritage, but we are all responsible for the way in which we choose or fail to take it into account in analyzing the world economic system, its injustices, and the need for change.[28]

Connecting the histories of the slave, peasant, and colonial era uprisings to the more recent developments of global warming, a line can be drawn to the inequality that climate change refugees experience today. 'The battle of equality will continue in the twenty-first century, basing itself chiefly on the memory of past struggles', Piketty claims.[29] Millions of people living in the most vulnerable regions affected by climate change are being forced to leave their now unfertile lands due to drought and water scarcity to seek refuge in detention camps or risk their lives to reach rich countries in the hope of finding better opportunities for survival. When able to break through the borders and barriers of Western nations, many refugees fall victim to their illegal status and are exploited as cheap labour on the farms and factories of Europe and America.

Creating the provision for global equality can only be achieved if wealthy nations transfer their excessive wealth and economic power to poorer countries. What is alarming and obvious in the 21st century is that global economic inequality is widening at a faster rate than ever before. The question facing humankind now is whether the world's wealthy nations are ready to make the necessary lifestyle changes for global equality as the basis in achieving global sustainability. A 2022 report by the World Inequality Database (WID) lists country-by-country income comparisons and overall wealth. Filled with in-depth statistics and graphs, the report calculates that while hundreds of millions of people have been lifted out of poverty, most notably in China, South-East Asia, Mexico, and parts of Africa, global inequality between countries and individuals has increased. The WID indicators are not surprising given the gross inequality of the tax laws for the super wealthy and global corporations that remain grossly in their favour. Taxing the rich and global corporations appropriately in combination with taxing the world's biggest carbon-polluting countries whose economies have reaped great wealth from using fossil fuels with scant regard for the damage caused is the only ethical way forward to secure global sustainability.[30]

If we are to agree that Western privilege has ridden on the back of colonial invasion, enslavement, and resource exploitation to feed the European industrialization of the 18th, 19th, and 20th centuries to the climate turbulence in the 21st century, then the most urgent task for the international community must be global economic equality. Fundamental to tackling climate change, the agreements made at the 2022 COP27 conference by rich Western nations to pay reparations to countries do not deal with the root causes of global inequality. As cited in Chapter 1, despite the agreement reached by countries to finance a 'loss and damage' Adaptation Fund amounting to US$100 billion a year, only US$10 billion has been committed. As Piketty argues, '[w]estern powers would be wise to take it [universal equality] seriously. If they persist in defending an obsolete hypercapitalist model, it is not at all certain they will succeed. The true alternative is democratic socialism, participatory and federalist, ecological and multicultural'.[31] How Piketty's global federalism might emerge remains an unknown where resistance to reparations from wealthy countries to former colonial countries and first nations peoples around the world still dominates the global agenda; nevertheless it is an essential step towards achieving global environmental recovery.

Economic equality is one way to close the gap of the disproportionate wealth, access to services such as education and health, freedom of movement, and the far-reaching racial, cultural, religious, and gender differences and discrimination. The journey towards human equality is not only an issue of wealth distribution but an ecological rediscovery of the natural world—a return that humankind must collectively embark upon to restore the earth's

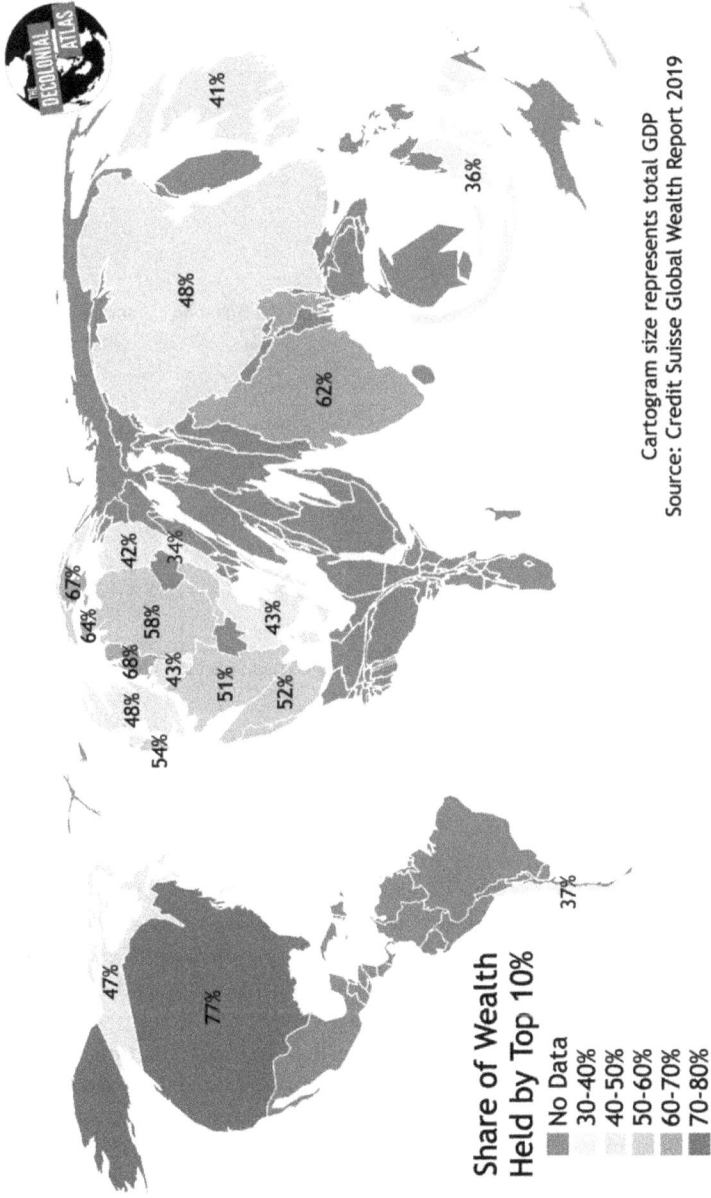

Figure 3.5 Share of wealth held by top 10%. Cartogram Jordan Engel, 2019.

Source: Image courtesy of The Decolonial Atlas.

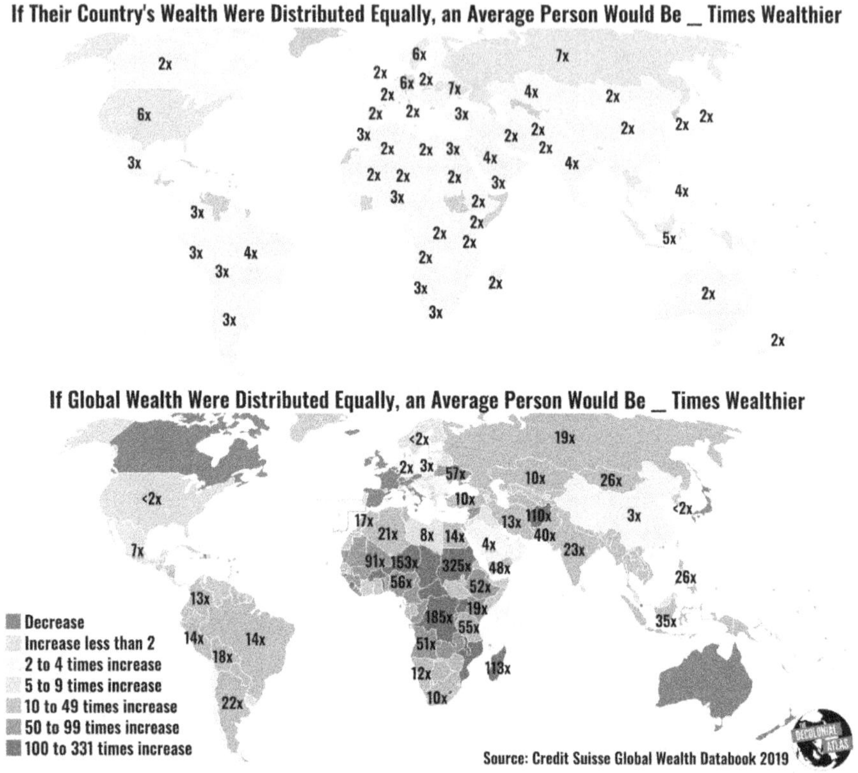

If Their Country's Wealth Were Distributed Equally, an Average Person Would Be _ Times Wealthier

If Global Wealth Were Distributed Equally, an Average Person Would Be _ Times Wealthier

Decrease
Increase less than 2
2 to 4 times increase
5 to 9 times increase
10 to 49 times increase
50 to 99 times increase
100 to 331 times increase

Source: Credit Suisse Global Wealth Databook 2019

Figure 3.6 Global wealth distributed equally. Cartogram Jordan Engel, 2019.

Source: Image courtesy of The Decolonial Atlas.

ecosystem. Some commitments to ecological restoration are undertaken, but only if they do not harm economic growth. Politics and capital foreground a system based on the desire and accumulation of wealth through the deployment and administration of labour and production. For global sustainability to be successful, the central position of politics and the economy that maintains world order would need to give way to the restoration of geological and ecological life forms of the earth as the real meaning of wealth. Centring ecological wealth and sending economic power and wealth to the periphery would involve a new engineering of human habitation on the earth on a scale not seen since the establishment of settlement. This re-engineered global society is no doubt the aim of sustainability. Redistributing the abundance of products and privileges that rich Western nations have enjoyed for centuries at the expense of poor countries would ensure global equality is entwined with global ecological sustainability. Overcoming key industrial

WHO'S RESPONSIBLE FOR CLIMATE REFUGEES?
Top 20 Countries[1] by...

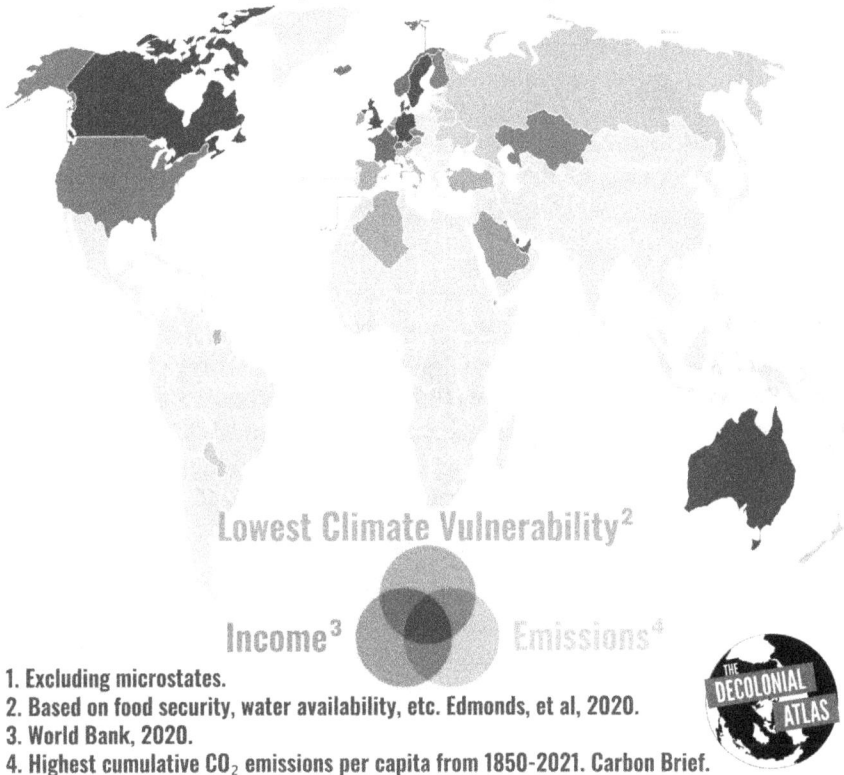

Lowest Climate Vulnerability[2]

Income[3]   Emissions[4]

1. Excluding microstates.
2. Based on food security, water availability, etc. Edmonds, et al, 2020.
3. World Bank, 2020.
4. Highest cumulative CO_2 emissions per capita from 1850-2021. Carbon Brief.

THE DECOLONIAL ATLAS

Figure 3.7 Who's responsible for climate refugees, 2021.

Source: Image courtesy of The Decolonial Atlas.

nations' objections to forming a new world order by removing their economic supremacy might prove an insurmountable objective. At the societal and national levels, this is more difficult given the prevalent use of national and patriotic rhetoric by right-leaning parties of rich Western nations intended to instil fear into their populations over the prospect of losing their wealth as much as being taken over by refugees at their borders. The siege mentality rhetoric by the rich over the poor will continue the social, racial, and economic divisions until mass human conflict and global civil collapse takes hold.

The short-term political life cycle of governments means that the latter often do not attend to the long-term ecological issues facing the planet.

Governments are concerned with the labour market, manufacturing base, trade imbalances, cost of living, and maintaining of public services provision such as pensions, health, and education. Governments are focused on implementing policies aimed at gaining favourable responses on the part of citizens rather than taking the risk of adopting unpopular policies that would diminish their chances for re-election. The focus of individual countries is more to preserve wealth and compete with countries through trade agreements than to band together and create a worldwide equitable and sustainable future Earth. For many countries, alleviating global inequality and poverty is not central to their political and economic aims. For example, initiatives such as switching from carbon-based energy production to renewables are said to put jobs at risk as fossil fuel companies lobby political parties to get support for their industry. The balance sheet is not a straightforward one: shifting from the continued use of fossil fuels to the rise of green energy means saving the planet from environmental destruction. Similarly, the contentious use of nuclear energy as a source of energy production, which presently accounts for 10% of global energy, is set to increase. Previously decommissioned reactors have reopened and 100 new plants that arc cither in the planning or construction stages will add to the existing 440 plants currently in use. All these measures are designed to maintain wealth through maintaining energy production, consumption, and jobs—not equity or equality.

Across the world, energy production is increasing to service the energy needs of an increasing global population. The systemic interweaving of politics, corporate power, nuclear, green, and fossil fuel use is enforced at varying levels on a country-by-country basis in a world experiencing the systematic dismantlement of ecologies, regions, and their people's ability to survive devastating climate events. Degradation of land and oceans is one clear result of climate change in relation to global inequality. Another significant effect is the decrease in global fresh water supplies. Unregulated water uses in agriculture and a culture of taking water for granted threaten people's survival especially in rural areas. Global fresh water consumption is far exceeding supplies, causing a major challenge to agriculture and furthering embedding inequality. Increasing populations around the world are suffering food and water scarcity, drought and famine due to climate change, while rich Western countries suffer least—a situation which the UN Secretary General Antonio Guterres decries as 'immoral' and 'untenable'.

The process of rich Western nations sharing their wealth for an equitable world must be viewed as an inevitability rather than an option to achieve a global sustainable future. When 'the global bottom 50% owns less than 1% of total wealth and the global top 10% nearly 82% of it', this challenges the morality by which rich Western nations and individuals presently live. The immorality of global wealth inequality is clear to see, yet the gap remains vast and the Western world's reluctance to fill it is evident. The capital system of wealth accumulation as a desirable goal uses a calculus to identify the wealth of others and this is taken as a guide for living on the earth. Initiatives such as The Giving Pledge, undertaken by billionaire philanthropists such as Bill and Melinda Gates, Warren Buffett, George Soros, Larry Ellison, and Bill Ackman are committed to giving away their fortunes to support programmes to combat climate change and global inequality, support food aid programmes, fund research into curable diseases and vaccinations, and provide micro-financing for small businesses. Such pledges relieve others from having to do the same and the number of billionaire philanthropists who have signed up to The Giving Pledge has in fact decreased over the last couple of years. Commitments quickly wear out, just as the impact of destabilizing and shocking news of devastating weather events rapidly become yesterday's news. Environmental collectives such as Fridays for Future, Extinction Rebellion, Just Stop Oil, and Last Generation have taken it upon themselves to take control through mass street protests and civil disobedience to bring about real change in raising awareness about climate change due to the ongoing use of fossil fuels. The actions in which these organizations are engaged, which often lead to arrests, fines, harassment, and branding as left-wing radicals, are in stark contrast to a far greater proportion of society who remain bystanders in the fight for climate change action. This constitutes another prong of social inequality.

Nationalists, economic protagonists, territorial protectionists, climate change deniers, and sceptics who place national interests before global welfare are engaged in a battle against the will of the people to take steps to combat the effects of global warming. The disproportionate levels of climate change devastation on the lives of hundreds of millions if not billions of people says that global inequality is not about to change. The fear, timidity, and outright refusal of rich Western nations to accept and engage with climate change victims is in itself not equitable, let alone sustainable. History has shown that the barriers to repel armies and people such as walls, guard dogs, and barbed wire can be scaled, blown up, or tunnelled under. Climate change is not bounded by walls, but its victims are. The Intergovernmental Panel on Climate Change (IPCC) 2022 AR6 Working Group II report says that 'approximately 3.4 billion people globally live in rural areas around the world, and many are highly vulnerable to climate change'.[32] A statement from the UN released in July 2021 reported:

> Overall, more than 2.3 billion people lacked year-round access to adequate food: this indicator—known as the prevalence of moderate or severe food insecurity—leapt in one year as much in as the preceding five combined. Gender inequality deepened: for every 10 food-insecure men, there were 11 food-insecure women in 2020.[33]

In another UN report published December 2021 by the UN Refugee Agency (UNHCR), it was reported that the global number of forcibly displaced people has 'passed 100 million people' and another report by ProPublica's Abrahm

Lustgarten in the *New York Times* of 23 July 2020 stated: 'Today 1% of the world is a barely livable zone. By 2070, that portion could go up to 19%. Billions of people call this land home. Where will they go'?[34] As previously cited in the introduction, these statistics make for chilling reading. To give a new statistic from research published in the journal *Nature Climate Change*, Annabelle Timsit and Sarah Kaplan in their article for the *Washington Post* of October 2021 write: 'At least 85 percent of the global population has experienced weather events made worse by climate change'.[35] As is clear, this number would seem to tell us that climate change is equitable, that it affects most of the earth's population. What it does not say is the number of distressed and vulnerable people whose lives are at risk and there it becomes clear that, though having no boundaries in the atmosphere, the huge impact of global warming on the ground is not equitable.

The earlier referenced IPCC AR6 Working Group III 2022 report, titled 'Mitigation of Climate Change', proposed five Illustrative Mitigation Pathways (IMPs) to halving global carbon emissions by 2030. The goals are as follows: (1) establish a global regime in rapid decarbonisation; (2) in the long term achieve net negative emissions by 2050 to meet 1.5 degrees global warming target; (3) rapid deployment and innovation on renewables; (4) reduce demand as rapid response to carbon emissions deduction; and (5) reduce global poverty, inequality, and broader environmental protections. All of these goals depend on financial backing and the will of governments and global institutions to carry them out. The steps required to achieve these goals are dependent on the interpretation and choices of individual countries to mitigate climate change impacts. In other words, it is based on the faith and a sense of global duty by each country, more so industrial countries, to reduce their greenhouse gas emissions. Both IPCC Working Groups II and III produced detailed documents about how governments, institutions, and corporations around the world can combine to establish various hybrid economic, industrial, and energy models to combat climate change and reduce global carbon emissions. These climate change documents involve government representatives, scientists, climate advocates, and reassuringly indigenous land care knowledges to formulate their global response. This bodes well with much of the hyperbole surrounding sustainability in the proliferation of climate conferences that combine reasons and arguments, actions, and agreements from participating countries and institutions. Putting the 'climate emergency' front and centre for governments, media and the general public is paramount in keeping the momentum to reduce global CO_2 emissions. Yet, once these conferences are disbanded, the implementation of agreed actions fall silent and the countries involved fall back on their governments and corporate economies of growth and profit.

Conflicting rational and irrational ideologies and programmes to tackle global warming are setting the future course of climate change and global inequality. While reasoned, rational approaches to combatting climate change are being less disputed, irrational climate change denial still persists across a broad political, corporate, and social spectrum in the face of overwhelming factual information and scientific evidence. Rational and irrational ideologies and programmes combine in the belief that human innovation and technology will offset climate catastrophe by reverse engineering the same technologies that have devastated the planet. In *Return to Reason*, Stephen Toulmin explores the dilemma that rationality correlates with our daily lives: 'As rational attitudes, optimism and pessimism cancel each other out. In the end, we can set aside dreams of eternal clarity, return to the World of Where and When, get back in touch with the experience of everyday life, and manage our lives and affairs a day at a time'.[36] The challenge in reducing climate change is managing 'everyday life' to everyday strategies in the long term to radically shift how humanity manages the earth's resources, energy production, and consumption. To achieve such changes requires the will of individuals to change their habits and collectively come together to force governments and corporations to change course and shift away from the dogma of continuous economic growth no matter the cost. The damaging consequences of entrenched global inequality have in some parts of the world resulted in a battlefield of life and death arising alongside the devastating effects of climate change. In the DRC, Central African Republic (CAR), and Central America, for instance, war, terrorism, and gang violence are a direct consequence of the combination of inequality and criminality fighting for control over resources. Financing global sustainability is therefore dependent on de-financing and de-arming militias in regional conflicts as well as containing global expenditure on the illegal sale of weapons. Where peace ensures the security of civil society, war and conflict ensure regional and continent destabilization and inequality. As the consequences of climate change continue to mount, devastating whole regions through increasing cycles of flood and drought, turning once fertile lands barren, and contaminating water supplies, reverting climate change becomes ever more insurmountable. We now accept that keeping global warming to 1.5 degrees by 2030 compared with pre-industrial levels is a pipedream in the face of the possibility of a 3 degree increase by 2080 and possibly even by 2050.

In her chapter 'Historical Ecology: Integrated Thinking at Multiple Temporal and Spatial Scales', Carole E. Crumley outlines how humans must respond to the challenges of climate change:

> The term *environment* must encompass the built environment, the cultural landscape, and nature wild and tame. The definition of *ecology* must include humans as a component of all ecosystems. The term *history* must include that of the Earth system as well as the social and physical past of our species.[37]

Speaking at the opening of the Petersberg Climate Dialogue Conference in Berlin on 18 July 2022, UN Secretary General Guterres gave another speech outlining the litany of natural disasters, stating: 'half of humanity is in the danger zone from floods, drought, extreme storms and wild fires'. At the time of his speech, heatwaves, wildfires, and droughts were ravishing parts of Southern France, Spain, and Portugal as a result of the highest temperatures ever recorded across Europe. The timing of Guterres' speech was aimed at calling attention to the ineffectual and incongruent global collective to tackle climate change, arguing that 'we are failing to work together as a multilateral community'. Guterres ended with a dire warning: 'we have a decision to make: collective action or collective suicide'. The ongoing terrestrial extractions by fossil fuel corporations, illegal logging of rainforests, the contamination of global water supplies are jeopardizing the lives of billions of people. Massive shifts in wealth inequality, opportunity, mobility, exchange, experience, and understanding between the world's urban and rural populations are fundamental to achieving global sustainability. The way humanity has inhabited the world is through the right to plunder the earth as it sees fit for its own progress. Humanity's estrangement from the natural environment set the course of the human psyche—an unfiltered comprehension that all living things are subject to human design.

The following chapter, 'Weathering Patterns', builds on Crumley's argument that '*ecology* must include humans as a component of all ecosystems' by exploring humans and ecology as a single comprehensible entity. For all of humanity's intelligence, its ability to return to the natural environment becoming wiser than before remains elusive. How humanity can reverse its estrangement, comprehend the natural environment not solely as an opportunity to profiteer from but as a pathway for humanity to return to the natural world as equals, is essentially the question of how humanity can save itself. Exploring the complexity of interconnecting systems of the animal and plant world offers the opportunity for humanity to reinvent itself not as animal or plant but as an integrated part of their system and could mark the start of human beings 'wearing our ecology'.

Notes

1 Lukáš Likavčan. *Introduction to Comparative Planetology*. Moscow: Strelka Press, 2022, p. 8.
2 For a recent report on the toxic effects of gas flaring, see BBC report: www.bbc.com/news/science-environment-63051458
3 An example of the widespread oil industry use of gas flaring—the burning of excess gas in oil extraction—can also be illustrated by Russia burning the equivalent of $US10 million dollars of gas that amounts to 9,000 tons of carbon being emitted into the atmosphere every day. Some of this amount is due to excess supplies from shutting down its gas pipeline to Europe as a result of the EU's support and military assistance to Ukraine to defend itself from Russia's invasion. For more information on gas flaring, see Global Gas Flaring Reduction Partnership

(GGFR) at: https://flaringventingregulations.worldbank.org/ and a report from the BBC: www.bbc.com/news/science-environment-63051458

4 The film *Under Poisoned Skies* directed by Jess Kelly, producer Owen Pinnell for BBC Arabic Investigations, 2022, can be viewed at: www.youtube.com/watch?v= gya0TXvoC88

5 See Likavčan, *Comparative Planetology*, p. 72.

6 See Martin Arboleda. *Planetary Mine: Territories of Extraction Under Late Capitalism.* London: Verso, 2020, p. 18.

7 Ibid., p. 29.

8 In Australia, the mining of iron ore alone increased by 31 million tons to 867 million tons in 2021, while coal increased to 75,428 million tons. For ore extraction, see www.minerals.org.au/minerals/ironore. For coal extraction, see www.ceicdata. com/en/indicator/australia/coal-production

9 For more information on Amazon rainforest mining, see Forests and Finance, Impact of Mining on the Brazilian Amazon: https://forestsandfinance.org/mining/ the-impact-of-mining-on-the-brazilian-amazon/

10 For a detailed report of mining in the DRC, see Cool Earth report 'Mining in the Democratic Republic of Congo' on its website www.coolearth.org/news/min ing-drc/

11 For more information concerning the palm oil industry in the Indonesian state of Borneo, go to the Eco Age website: https://eco-age.com/resources/the-truth-beh ind-the-palm-oil-industry-through-the-eyes-of-a-bornean/

12 See United Nations Convention to Combat Desertification, *Global Land Outlook*, 2nd edition. Bonn: UNCCD, 2022, p. 10. https://insideclimatenews.org/news/ 27042022/agriculture-land-report/

13 See Likavčan, *Comparative Planetology*, p. 17.

14 Ibid., p. 18.

15 See W. B. Higman. *Flatness.* London: Reaktion Books, 2017, p. 220.

16 See Marcus Vitruvius, *De Architectura*, 'Ten Books on Architecture', translated by Morris H. Morgan. Cambridge, MA: Harvard University Press, 1914.

17 See Glenn A. Mazis. *Humans, Animals, Machines: Blurring Boundaries.* Albany, NY: SUNY Press, 2008, p. 21.

18 See Nnimmo Bassey. *To Cook a Continent: Destructive Extraction and the Climate Crisis in Africa.* Cape Town: Pambazuka Press, 2012, p. 7.

19 See Vandana Shiva. *The Green Revolution: Third World Agriculture, Ecology and Politics.* London: Zed Books, 1993, p. 14.

20 See John Bellamy Foster and Brett Clark. *The Robbery of Nature.* New York: Monthly Review Press, 2020, p. 7.

21 Ibid., p. 310.

22 See Julia Kristeva and Anna Smith. *Readings of Exile and Estrangement.* London: Palgrave Macmillan, 1996, p. 3.

23 For a concise breakdown of the African continent's CO_2 emissions over the last 20 years, go to the Statista website: www.statista.com/statistics/1287508/africa-share-in-global-co2-emissions/#main-content

24 See Nancy Harris et al. 'Global Maps of Twenty-First Century Forest Carbon Fluxes'. *Nature Climate Change*, 11(2021), 234–240; *The Economist*: www. economist.com/interactive/graphic-detail/2022/05/21/the-brazilian-amazon-has-been-a-net-carbon-emitter-since-2016

25 See James Lovelock. *Gaia: A New Look at Life on Earth*. Oxford: Oxford University Press, 2000, p. 10.
26 Lovelock, who died in 2022 at the age of 103, equates Gaia to the human body stating:

> the relationship between biosphere and Gaia is like that between your body and you. In other words, we are inseparable from the biosphere as our bodies are inseparable from our mind. Yet, the common fault of humankind is to precisely form this separation and if we can separate ourselves from our bodies then we can separate or disconnect ourselves from the biosphere.
>
> (Lovelock, *Gaia*, p. xii)

27 Thomas Piketty. *A Brief History of Equality*, translated by Steven Rendall. Cambridge, MA: The Belknap Press of Harvard University Press, 2022, p. 26.
28 Ibid., p. 49.
29 Ibid., p. 226.
30 In the abundance of information produced by the World Inequality Database (WID) 2022 report, some statistics stand out by their simple comparative analysis. On page 27 of the 234-page document, the authors make clear how much global wealth is disproportionately distributed:

> Global wealth appears to be even more unequally distributed than global income. The poorest half of the world population owns just 2% of total net wealth, whereas the richest half owns 98% of all the wealth on earth. The bottom 50% owns, on average, €2,900 of assets (typically in the form of land, housing, deposits or cash). Between the richest half of the global population, the middle 40% owns just 22% of total wealth (on average €40,900 per adult) and the top 10% owns 76% (i.e., €550,900 per adult, on average, including a large share of financial wealth such as stocks and bonds).
>
> (WID 2022 report by World Inequality Lab, coordinators/authors Lucas Chancel (lead author), Thomas Piketty, Emmanuel Saez, Gabriel Zucman, United Nations Development Program, p. 27)

31 Piketty loosely discusses how to reach global equality and counter the rise of climate change, claiming

> All the transformations discussed in this book, whether the development of the welfare state, progressive taxation, participatory socialism, electoral and educational equality, or the exit from neocolonialism, will occur only if they are accompanied by strong mobilizations and power relationships. There is nothing surprising about that: in the past, it has always been struggles and collective movements that have made it possible to replace the old structures with new institutions. Nothing prevents us from imagining peaceful developments, supported by new social and political movements that have succeeded in mobilizing a large majority of voters and rising to power on the basis of platforms proposing ambitious transformations.
>
> (Piketty, Brief History of Equality, pp. 226–227)

32 See Intergovernmental Panel on Climate Change (IPCC), Adaption and Vulnerability, AR6 WGII Summary for Policy Makers Impact, March 2022, Switzerland, p. 45.

33 For more information on the connections between gender and food security, see www.who.int/news/item/12-07-2021-un-report-pandemic-year-marked-by-spike-in-world-hunger

34 To read Abrahm Lustgarten's full article of 23 July 2020 in the *New York Times*, see www.nytimes.com/interactive/2020/07/23/magazine/climate-migration.html

35 For more information, go to www.washingtonpost.com/climate-environment/2021/10/11/85-percent-population-climate-impacts/

36 Toulmin's reference to rationality is in part directed at the scientific method of philosophy of human behaviour that slides across the spectrum of the rational and irrational:

> If the idea of rationality is problematic, that of 'irrationality' is even more difficult. If formal logic were truly the science of rationality, we would expect irrationality to show itself in errors of formal reasoning. How this connects to human behaviour and the ill effects of climate change is the irrationality of humanity's persistent destruction of the earth's natural world for the rationality of extending the imbalance that humans are impacting on the earth. Human behaviour and political will are proving hard to change especially when humans have spent centuries in the belief that they can dominate the earth.
>
> (Stephen Toulmin, *Return to Reason*. Cambridge, MA: Harvard University Press, 2003, p. 213)

37 See Alf Hornborg and Carole L. Crumley eds. *The World System and the Earth System*. Walnut Creek, CA: Left Coast Press, 2006, p. 27.

Bibliography

Arboleda, Martin. *Planetary Mine: Territories of Extraction Under Late Capitalism.* London: Verso, 2020.

Bassey, Nnimmo. *To Cook a Continent: Destructive Extraction and the Climate Crisis in Africa.* Cape Town: Pambazuka Press, 2012.

Bellamy Foster, John, and Brett Clark. *The Robbery of Nature.* New York: Monthly Review Press, 2020.

Higman, W. B. *Flatness.* London: Reaktion Books, 2017.

Hornborg, Alf, and Carole L. Crumley, eds. *The World System and the Earth System.* Walnut Creek, CA: Left Coast Press, 2006.

Kristeva, Julia, and Anna Smith. *Readings of Exile and Estrangement.* London: Palgrave Macmillan, 1996.

Likavčan, Lukáš. *Comparative Planetology.* Moscow: Strelka Press, 2022.

Lovelock, James. *Gaia: A New Look at Life on Earth.* Oxford: Oxford University Press, 2000.

Mazis, Glenn A. *Humans, Animals, Machines: Blurring Boundaries.* Albany, NY: SUNY Press, 2008.

Piketty, Thomas. *A Brief History of Equality.* Translated by Steven Rendall. Cambridge, MA: The Belknap Press of Harvard University Press, 2022.

Shiva, Vandana. *The Green Revolution: Third World Agriculture, Ecology and Politics*. London: Zed Books Ltd., 1993.

Toulmin, Stephen. *Return to Reason*. Cambridge, MA: Harvard University Press, 2003.

Vitruvius, Marcus. *De Architectura* 'Ten Books on Architecture'. Translated by Morris H. Morgan. Cambridge MA: Harvard University Press, 1914.

4   Weathering Patterns
Entering the Biosphere

Animal—Ecological Comprehension

According to biblical accounts, God created the earth and all living things in six days. Resting on the seventh he looked back on his creation to see a great diversity of animals, plants, and a couple of humans each tasked with multiplying their species. Sometime later, looking down on the earth to great diversity of species he created, he saw humans worshipping pagan icons and sought their punishment. Ready to sacrifice it all, he called upon a man called Noah who had not deserted him to build an ark and load it with animals, plants, and seeds. When Noah completed the task, God unleashed torrential rains to flood the earth and end all remaining human, animal, and plant life except for Noah, his family, and the animals and plants he had gathered in the ark. What God was able to do by flooding the earth was to showcase his ability for creation as much as destruction, and thousands of years later, humans would likewise be capable of doing the same.

The creative innovations that characterized the immense advancements in 20th-century technology made it possible for humans to carry out their capacity for destruction of the earth and of itself. The biblical story of God's mass extinction as an act of vengeance against the pagan worshippers has, in today's world, flipped. The earth is now on the rampage, avenging what rampant industrial growth, consumption, and the destructive forces of burning oil, gas, and coal have done to its ecological system. Now humans have become God-like, destroying the earth with the same religious zeal to physically alter its surface and decimate animal and plant life. Since the inception of institutional belief, religion has played a prominent role in how humans interface with the earth. Under God, animal and plant life paid a high price for what he saw as human infidelity but as with all epic stories, cracks soon begin to appear when you pull the narrative apart. The German zoologist, naturalist, and eugenicist Ernst Haeckel in his book, *The History of Creation, or, the Development of Earth and Its Inhabitants by the Action of Natural Causes*, Vol. 1, published in 1876, raised some questions that unsettled the accepted narrative of Noah's ark, for when the floods had subsided and Noah

DOI: 10.4324/9781003382515-5

released the animals and planted the plants and seeds to repopulate the earth, there would be problems.

> As to the simultaneous origin of all individuals of each species from one pair...[it] is clearly quite untenable...[T]he few animals of prey would have sufficed to have utterly demolished all the herbivorous animals, as the herbivorous animals must have destroyed the few individuals of the different plant species of plants.[1]

Lions, tigers, hyenas, crocodiles, snakes, lizards, spiders, etc. would, by natural instinct, have sought out their prey of deer, gazelle, zebra, rabbit, mice, butterflies, and insects. In a short space of time, only apex animals would be left before they would have to turn on themselves until there was nothing left for the last lion or tiger to eat, causing them to die from starvation.

Fraught with discrepancies, the Christian God's destruction of the innocent plants and animals is not necessarily transferable to other religious beliefs where animal and plant life is venerated. Islam, for example, considers nature as a sign and symbol of God (Allah) on Earth. In the *Quran*, there are approximately 300 references decreeing a sustainable coexistence connecting Islam and the natural world. It asks its followers to 'not commit abuse on the earth, spreading corruption'. 'It also decrees the planting of trees for the propagation of fruits they bear as an example of the provision Allah provides'.

Quran, 2:60 مَّشْرَبَهُمْ ۖ كُلُوا۟ وَٱشْرَبُوا۟ مِن رِّزْقِ ٱللَّهِ وَلَا تَعْثَوْا۟ فِى ٱلْأَرْضِ مُفْسِدِينَ ٦٠

And He is the One Who sends down rain from the sky—causing all kinds of plants to grow—producing green stalks from which We bring forth clustered grain. And from palm trees come clusters of dates hanging within reach. [There are] also gardens of grapevines, olives, and pomegranates, similar [in shape] but dissimilar [in taste]. Look at their fruit as it yields and ripens! Indeed, in these are signs for people who believe.[2]

وَهُوَ ٱلَّذِىٓ أَنزَلَ مِنَ ٱلسَّمَآءِ مَآءً فَأَخْرَجْنَا بِهِۦ نَبَاتَ كُلِّ
شَىْءٍ فَأَخْرَجْنَا مِنْهُ خَضِرًا نُّخْرِجُ مِنْهُ حَبًّا مُّتَرَاكِبًا
وَمِنَ ٱلنَّخْلِ مِن طَلْعِهَا قِنْوَانٌ دَانِيَةٌ وَجَنَّٰتٍ مِّنْ أَعْنَابٍ
Al-An'am:99 | 6:9 وَٱلزَّيْتُونَ وَٱلرُّمَّانَ مُشْتَبِهًا وَغَيْرَ مُتَشَٰبِهٍ ٱنظُرُوٓا۟ إِلَىٰ ثَمَرِهِۦٓ إِذَآ أَثْمَرَ وَيَنْعِهِۦ إِنَّ فِى ذَٰلِكُمْ لَءَايَٰتٍ لِّقَوْمٍ يُؤْمِنُونَ ٩٩

Unlike Christianity, Islam forbids the visual depiction of Allah who is 'all present' in nature as the consciousness of Him. Through architecture and building, the design of Islamic mosques weaves intricate geometric complexity to illustrate the majesty of Allah's universal knowledge in accordance with nature. Christian religious architecture, on the other hand, is designed for ascendency in the form of the spire piercing the heavens and as ground plan in the crucifixion of Jesus on the cross. Jesus's soul and outstretched arms are represented as the alter and side chapels as the place of his devotees and the torso and legs as the place of his followers—the congregation arranged in pews. Islam's veneration of nature and Christianity's destruction are clearly reflected in their architectural religious representation and relationships with the earth. Where Christianity is founded on revenging idolators of false gods and debauchery and involves the annihilation of them and the natural world, Islam embraces nature in a simultaneous belief in Allah. In terms of humanity and the earth, these differences would have a profound impact and immense repercussions throughout the world. Christian doctrine advocated the belief that nature and all its resources exist for the sole provision of humanity, and European Christian colonialists imposed this orthodoxy on the civilizations they invaded, enslaving first nations peoples and plundering their lands for profit.

Apart from Christianity's and Islam's relationships to nature, other religions formed their spiritual connections with the natural world. Hindu belief ties physical interconnectivity to the earth—water, the earth, fire, and air are seen as the primal energy forces to give life and form between humans and nature. The universal principle of Brahman, whom Hindus revere as the ultimate deity, encompasses male and female spiritual form as well as animal form to coexist in physical iconic partnership as Islam and Christianity do in building. Buddhist belief is concerned with the empowerment of self-enlightenment entwined with the divinity of all living things in the natural world. This includes not only animals and plants but also the geography of rivers and mountains and the geology of rocks all of which are composed of Buddha's natural enlightenment. Taoism follows a similar path to Hinduism and Buddhism where nature is at the centre of human universal connectivity. Complex and multifarious indigenous beliefs are intertwined with nature on Earth and the cosmos. They too are physical as can be seen in Australian indigenous people's human–animal–spiritual ancestors who burrow underground to forge geography to *become* nature's appearance.

Distinct from religious connective creations between humans and the earth is the life of animal and plants which are physically set to environmental adaptation and transmutation for their evolution. The English naturalist, Charles Darwin conceived of his theory of evolution or Transmutation Theory in his seminal work, *On the Origin of Species* published in 1859. Darwin's theory of adaptation or transmutation founded his concept of natural selection whereby the physical characteristics of the earth's natural environments accounted for the biodiversity of animals and plant species. Mammalia (whales, rats, apes, pigs, sheep, lions, humans etc.), Reptilia (Komodo, turtles, crocodiles etc.), Aves (birds), Invertebrates (octopus, cuttlefish, snakes, worms, spiders etc.), Amphibia (frogs, toads, salamanders etc.), Fish (aquatic, craniate, gill-bearing animals) to name a few attest to species diversification in relation to their environments. Human evolution did not evolve in the same way to produce adaptive species diversity as animals and plants; instead humans developed their cognitive ability to harness the natural world to their needs. Over hundreds of thousands of years, humans and their ancient ancestors competed with animals for survival that ended with the founding of agriculture and the domestication of animals. As humans increased their control over animals, they also advanced a perverse relationship that can be seen in how pigs today are bred with four more ribs, cows are genetically engineered to double their daily milk production, salmon are genetically engineered to mature rapidly, and chickens are processed within six weeks of hatching to the dinner table. The growth of industrial-scale animal farming for human consumption has transpired to add 14% of greenhouse gas emissions from livestock methane. Other than ceasing to eat animal products, plans to offset these emissions have created a different set of directions concerning animal modification, taking human interference in animal life to new levels such as genetically engineering sheep and cows to produce less methane from burping and flatus. These early-stage scientific experimentations point not to a technological focus on biological modification but rather to a complete overhaul of animal consumption. Re-engineering animals rather than changing human behaviour reveals the subservient position that humans still assign to animals.

Relations between humans and animals vary but nowhere is animal perversity more profound than in rich Western nations where cats and dogs far better than humans in many poor countries. In 2019, the average yearly amount Australians spent on their dogs was A$1,475 and on cats A$1,029. Americans spent US$1,678 on dogs and US$1,116 on cats amounting to $52 billion in total and in the United Kingdom £2,880 was spent on dogs and £1,200 on cats. By comparison, 10% of the world's population (800 million people) live on less than $2 a day or US$730 a year. The disparity that exists between large populations living in poverty and the money spent on pets reveals a value system that is difficult to comprehend and yet it is widely accepted if not embraced. Humans' idolization of their pets differs from God's pagan idolators and remains indifferent to the millions of animals they consume. Hunted, killed, and eaten or glorified as pets, humans maintain an ambiguous, self-serving, and destructive relationship with animals.

The early human shaping of their environment began in earnest with the invention of tools. The advancements in tool design and application shaped civilization and characterized humanity's 'geophilic' love of stone forging an uncountable range of new possibilities. Human geological 'friending' of stone would convey its historical passage for 'rock archives it all'. In *Stone: An Ecology of the Inhuman*, Jeffrey Cohen assigns 'geophilia' to a human relationship with the earth without indifference to human experience. Rock is neither sedentary nor solitary, Cohen informs us, 'but networked matter, full of movement and connection'.[3] Cohen views stone as the caretaker of human physical and mental progression.

Humans walk upright over earth because the mineral long ago infiltrated animal life to become a partner in mobility. Vertebral bone is the architect of motion, the stone around which the flesh arranges itself to slither, run, swim, fly. Had the organic not craved durable calcium as shield and conveyor, numerous types of sedimentary rock would never have arrived.

Cohen's stone archaeology not only archives human life on the earth but is also predictive concerning its future. 'Because the scale of its unfolding is immense, geophilia entwines the modern and the ancient, the contemporary and the medieval, the primordial with expansive futurity'.[4] In *Theory of the Earth*, Thomas Nail refers to animals not only as a physical presence but also a co-conscious embodiment of the earth.

> The emergence and proliferation of animals on the earth was the source of a radical new regime of elastic motion, characterized by the ability of living matter to expand, contract, stretch, and oscillate back and forth to a degree never before seen on the earth... Their existence changed the entire bio-geo-chemo-atmo-sphere. The earth, in other words, *became animal*.[5]

As with Cohen, Nail's universal osmosis of particle connectivity is further embodied in their shared history: 'Every body (mineral, atmospheric, vegetal, and animal) carries in itself the entire history of the earth that led up to its existence and is pregnant with the matter of future bodies'.[6]

In 'Why Look At Animals?' from *About Looking*, John Berger begins with a solemn account of the nature of human destruction. 'The 19th century, in western Europe and North America, saw the beginning of a process, today being complete by 20th century corporate capitalism, by which every tradition which has previously mediated between man and nature was broken'.[7] The hunting and killing, clearing of land, and felling of forests have allowed humans to believe that with stone, bow and arrow, dart, spear, and later gun, plough, and machine, they can progressively separate themselves from plant and animal life. Berger uses the example of the domestic cow whose purpose was not solely the production of milk and meat but to perform a magical and sacrificial ceremonial function. 'Animals first entered the imagination as messengers and promises', he tells us. Early human veneration and sacrifice of animals to cultural and religious ceremonies denoted the constructed discordance and resemblances between them—it was apparently what the poor pagans engaged in that God so disliked as to 'kill them all'. Berger talks of animal and human similarities, for as with humans, not every animal is the same and they are united; 'sentient', 'mortal', and 'born'. He describes how animals scrutinize man 'across the narrow abyss of non-comprehension. This is why the man can surprise the animal' and the animal can 'surprise the man' for he too 'is looking across a similar, but not identical, abyss of non-comprehension'. Berger describes man's abyss as 'looking across ignorance and fear. And so, when he is *being seen* by the animal, he is being seen as his surroundings are seen by him'. Berger is concerned with the differences in humans' and animals' perceptions of each other rather than the existence of the abyss separating them in nature. Berger writes that the 'first subject matter for painting was animal' and it is likely that 'the first paint was animal blood', suggesting 'it is not unreasonable to suppose that the first metaphor was animal'.[8]

The abyss of non-comprehension between humans and animals expresses the fragile relationships that bind as much as separate them. In *Thinking Animals: Animals and the Development of Human Intelligence*, Paul Shepard suggests that early human recording of animals in cave rock art 'encode[s] a dual system in which the ecology of animals parallels the society of humans... The animals to which they refer live in an orderly world, a visible ecosystem'.[9] As human evolution gathered pace, this collective eco-system would no longer be mirrored and the human's view of animals would become domineering.

> For us the human will—the same imperious will that now creates every-thing around us—is our god. To it everything seems to come from within. The world is only a stage where we act out those scenarios that suit us. How could the ambience of nature be part of our physiology? How could ecology have anything to do with mind in such a view?[10]

Shepard points to the dominance of human perception towards the natural world as an attitude that led to the downfall of a shared ecology between humans and animals. To understand the route of this downfall, one need only look at the present environmental condition concerning the climate crisis humanity now faces; to avert further environmental freefall would mean for humanity to regain the ecology it once shared with animals.

The history of humanity can be characterized by its progressive, some would say impressive, dominance over nature. Given its perception of super-iority, the effects of climate change raise questions about how humans might secure the survival of the natural world as the premise to their own survival. Ensuring a mutual survival would require an aptitude to break the duplicity concerning human relations to the natural world for new coexistent relations in the future. In terms of animal relations, this would mean recreating a dual ecological human and animal community. 'An intrinsic part of human self-consciousness requires animal images, and no amount of urban or techno-logical glazing defeats it', Shepard notes. 'The ecology of creatures is the model for the "society" of abstract ideas', which suggests a way for humanity to undo its linear formation and relinquish its central position in relation to the earth to a more abstract one concerning the role that humans play. 'Any mix of plant and animal species works out its reciprocities over time and adjusts to changes', Shepard claims, which prompts the question: why not cultivate a similar reciprocity between humans, animals, and plants?.[11] To turn human consciousness from its historical roots and rethink its relations with the natural world is to subsume the ecology of animal and plant life to their own. 'Every human mind is a product of its ecology—the same ecology', Shepard informs us; '[n]othing that evolves persists unless sustained by those same creative forces. Like a ball at the top of a fountain, the human head pivots on its animal backbone, the mind a turning knot of thought and dream on the end of a liquid spear of living animals'.[12]

In another book, *The Others: How Animals Made Us Human*, Shepard suggests early human life was indistinguishable from animal life. 'The digging stick may have been the first human tool, opening the earth to reveal an inside like that of an animal or human body'.[13] Outwardly, humans could be animal; we look like them and imitate their behaviour, skin, and mask. 'The human mind', he explains, 'is the result of a long series of interactions with other animals'.

> The mind is inseparable from the brain, which evolved among our primate ancestors as part of an ecological heritage. That heritage began with life on the ground, continued in the trees, and millions of years later came back to the ground. This upstairs–downstairs legacy, arboreal and terrestrial, is not unique to our descent but is widespread among monkeys and apes, so that to understand our kind of consciousness—higher-level thinking, artistic expression, and abstraction—requires some further explanation and is linked to our perception of animals in a roundabout way.[14]

To understand human and animal relations to the natural world can be seen in how each adapted and flourished. In the Darwinian conception, animals adapt to the surroundings forming their transmutation, whereas humans utilize their evolution through the expansion of the mind by forming concepts and their ability for invention to build relations to their surroundings. As the ultimate expression of the human capacity for invention, the conceiving and building of cities were to become in design and material terms the recomposition of nature. Where animal comprehension of nature is inseparable from their existence, humans forged their environment to comprehend themselves as separate from and outside their surroundings. 'To be urban—to live in mass society at a distance from wild diversity—is to share a heightened angst about the pronominal enigma: the identity of I, we, you, it, and they', Shepard claims. 'As if to deny our poverty of wild things, we declare a cultural superiority over such "primitive" reference...Our schizoid alienation from the animals has led us to project the frightening confusion of our urban grayness upon them'.[15]

Where cities illustrate human comprehension by building-out the natural world, it is not surprising that humans apply a similar condition to shaping and 'unbuilding' it. For humanity to regain its ecological heritage, it must reformulate its relations to the natural world, to animals and plants, and in doing so become *their* ecology as against constructing their own. Returning to Berger's animal and human comprehension of the abyss: 'when he is *being seen* by the animal, he is being seen as his surroundings are seen by him'. For humans to regain ecology will mean bridging the abyss between them and animals. As co-sentient beings, humans relate to animals for we see ourselves in them—even though this connection is not reflected by the animal. 'What is true for the human species, that no two individuals are alike...is equally true for all other species of animals and plants', Ernst Mayr writes in *Animal Species and Evolution*.[16] Humans have applied their superior intelligence over all living things as a means of separable comprehension. To be human is not to identify as animal even though humans are animal-like. To be human is to stand aside from animals by constructing their isolation from the natural world. Animals on the other hand remain naturally inclusive, even between predator and prey where a cooperative balance of needs exists.

Humans are bipedal as are birds. Hoof, paw, claw, and flipper animals are quadrupedal. Some animals such as kangaroos are bipedal, quadrupedal, and pentapedal if you include the tail which acts as a fifth leg. Animals such as octopuses and squid (cephalopods) and crustaceans (molluscs) are octopedal and decapoda such as lobsters, crabs, and shrimps. Spiders and insects (arthropods) such as centipedes and millipedes are multipedal. As part of the evolutionary physiology of living organisms, humans pulled apart from their relations to animals as their intelligence developed and began to walk upright. For humans to rethink their relationships to animals sounds bizarre given their immense differences now but it is a necessary part of rebuilding the earth's devastated ecologies. Paraphrasing Berger and Shepard, Laurence Simmons in his book *Knowing Animals* notes that 'the animal exists in the abyss of a particularly differential otherness. Different from the sameness of another human, yet also different from the incommensurability of an inanimate object'.[17] Human indifference to animal and plant life has no doubt led to the devastation of environments and the extinction of species most notably during the 20th century, and the roots of climate change stem from humanity's comprehension of dominance over the natural world. In his chapter, 'Extinction Stories: Performing Absence(s)' from the same book, Ricardo De Vos writes: '[o]ur current notions of extinction are shaped both by the knowledge that more than 99 percent of all known animal and plant species are now extinct, and that we are currently living in a time of mass extinction'.[18] It is hard to fathom such a figure but considering 150–200 animal and plant species become extinct every day, it is extremely confronting to live with this actuality in one's lifetime. De Vos points to the era of European colonialism, which 'exacerbated the current pattern of mass extinction, and by the end of the nineteenth century, even the islands most remote from Europe had been subjected to the effects of sailors, scientists and settlers'.[19] This can be seen in Australia where the introduction of dogs, cats, pigs, buffalos, horses, camels, and cane toads in the 18th, 19th, and 20th centuries led to the decimation of native animal and plant life species and ravaged fragile landscapes.[20]

In *Anthropocentrism—Humans, Animals, Environments*, Sabrina Tonutti refers to the tower of cultural superiority humans have deployed in separating themselves from the animal world.

Descriptions and representations of the place of humans in nature are often conveyed by evocative images and metaphors: humanity has been portrayed as an 'island', the difference between the nature of animals and human culture as a 'rubicon', and man described as an 'empty container' filled with culture.

'Despite lacking in instincts and biological apparati', Tonutti continues, 'humans were seen as completed by culture, and it is this capacity that would emancipate them from the constraints of nature'.[21] Yet, human dominance and cultural superiority over nature, especially over animals, is so fragile that even when verifiable claims showing chimpanzees, apes, and monkeys using tools in the aid of acquiring food sent shockwaves among the world's ethologists (the study of animal behaviour), such claims were disputed. The prior accepted position in ethological thought was that humans were a distinct species of unrivalled intelligence such that any direct connections between them and chimpanzees, apes, and monkeys were scientifically absurd. Tonutti lists the work of ethologists such as Jane Goodall's studies of chimpanzees' intelligent use of tools. Though shunned for years by ethologists, Goodall's research and later Dian Fossey's work on the family structures of gorillas shone a light on their social and familial proximity to humans, which would eventually come to be universally accepted.

The social structures between animals and humans are easy to find even in the largest and most unlikely human appearance of animals. Elephant society is based on familial relations that flow across the herd to formulate a collective shared responsibility between the old and the young. Rats, on the other hand, coexist within a set patriarchal structure. Worker rats serve their sub-leaders and the sub-leaders serve the overall leader—the king rat. Communication and decisions between these sets of levels from king rat down to the worker rat are worked through these exchanges of command. Ants and bees structure their societies via matriarchal governance where the queen, the producer, and guardian of reproduction is served by worker ants and bees in maintaining the colony and hive. As with the elephant, king rat, queen bee and ant, their present and future survival is held in their DNA and if their herd, colony, or hive are attacked, their safety is paramount for the future survival of the herd or colony. Besides similar familial structures, human society works in far more complex structures if you consider the institutions of government, capital, religion, and culture administered in acceptable, disproportional, and inequitable systems and in varying degrees of wealth, gender, and racial bias.

In her book, *What It Means to Be Human*, Joanna Bourke takes the surface of the Möbius strip as an analogy to express the pathway of human and animal nature. A single spiral surface without beginning or end, inside or outside, the Möbius strip contrasts the linear trajectory humans have taken to separate themselves from the environment and from each other, animals, and plants. 'The boundaries of the human and the animal turn out to be as entwined and indistinguishable as the inner and outer sides of a Möbius strip', explains Bourke. It 'embodies the roller-coaster ride of life, or *zoe*. Most usefully for us, it deconstructs the human versus animal dilemma'.[22] Human ancestry has shared an extensive evolutionary timeframe of gestation with animal life forms in its physiological formation over millions of years. Bourke suggests that in order to move beyond this evolutionary history of human and animal distinction, a new fluidity regarding identity needs to move between them.

> The concept can be simply stated: face to face with the fundamental fluidity in definitions of human/animal (the twistings and turnings of that Möbius strip), we must move beyond comparisons based on similarities and dissimilarities and inject instability and indeterminacy into our discussions. The advantage of thinking in terms of the Möbius strip is that it encourages a celebration of difference and uniqueness.[23]

Bourke takes the view that 'the image of the Möbius strip provides a way of challenging tyrannical dichotomies such as biology/culture, animal/human, colonizer/ed, and fe/male'.[24] Further explored in Chapter 6, the Möbius strip offers a way of thinking how human society can be reconfigured to coexist on the same twisting single surface.

The first molecular beginnings of the human life form were established by their earliest divergence from fish approximately 360 million years ago. A clearer sign began to emerge when humans and apes began to diverge (as distinct from evolve) from a shared distant relative approximately 25 million years ago and an even clearer sign came when humans and apes fully separated approximately 5–7 million years ago. During most of this time, humans processed their animality with other animals, plant life, and the weather. Coming 'out-of-animal', humans were able to pit their larger brains and capacity for invention to rule over animals. In the rise of animal cultivation through domestication, humans wielded power over animals enhanced by physical constructions to entrap them. Wild animals, by contrast, continued to live according to their instincts conditioned by their environment and natural perception, being wary of humans and their ability to threaten, trap, and kill them. Taking these two distinctions into consideration, it is fair to assume that humans not only terminated their animality to increase their distinction but also designed their separation to formulate how to live on the earth. 'Vast in its analytic and inventive power', David Abram writes in his book *Becoming Animal: An Earthly Cosmology*, 'modern humanity is crippled by a fear of its own animality, and of the animate earth that sustains us'.[25] In reference to Charles Darwin's Theory of Descent or Transmutation Theory concept of evolution, Abram points to the animality humans have sought to escape '[b]y renouncing our animal embodiment—by pretending to be disembodied minds looking *at* nature without being *situated* within it'.[26] Mirroring the thoughts of Berger and Shepard on the human–animal physical comprehension and inner desire for acknowledgement, Abram argues that the human condition is precipitated on the expulsion of its animality: 'we've convinced ourselves that the land around us—generously abundant at some moments, ferocious and unforgiving at others—is really a set of sheer facts; we hold ourselves apart from the world in order to subdue its wildness'.[27]

In *What Animals Teach Us about Politics*, Brian Massumi charts animality in humans as the continuum 'of the human to the animal'. 'To think the human is to think the animal, and to think the animal is to think instinct. Would it even be possible to conceive of an animal without instinct?' he asks.[28] Is it possible for humans to regain their animality given their destructive animal history? The answer to this would have to be a resounding no. By separating out their animality, humans have shattered any vestigial ecology that might display their animality. As with Berger's, Shepard's, and Abram's ideas regarding human apprehension of animal identity, Massumi points to the rejection that humans constructed to remove their connection to animals. 'Even in the most intimate and humanly ordered situations in which animals frequent humans, in the role of companion animals or in animal husbandry, they never imitate the human. They relate to them. In this sense, the identification only goes one way'.[29] Yet, human separation from its animality can be reclaimed, according to Abrams.

> The theory of evolution by natural selection made evident that we, too, are animals, created not by an external divinity but evolved from a group of primate ancestors, and more anciently from a lineage of small mammals, themselves derived, much earlier, from an ancestral line of fish long immersed in the ocean depths.[30]

Working through the positions of Abram, Massumi, Shepard, and Berger, in terms of the relation between humans and animals, we can wonder how humans and animals would adapt to new climatic conditions as a result of global warming. The human evolutionary trail from fish–ape–human formulated early human connections with the natural world and reconnecting its ability for adaptation concerning climate change, now, depends on the divergence of the human race to obtain its assimilation with animal and plant life.

How does this focus on looking at animals fit within the book's project on sustainability? Berger and Shepard are clear in their formulation that humans look at animals to learn about themselves so as to make clear the distinction between the two. The massive effects of climate change are causing humans to review their unbalanced perception with the natural world and consider how they might overturn their past practices in respect of the earth, including animals. Climate change has brought into sharp focus human existence on Earth to foresee humanity's own extinction as it witnesses animal extinction. It has become clear that human impact on the world—the devastation of the natural environment—is at the same time the devastation of human life. Humans now understand that the extinction of animal and plant life is, as UN Secretary General Antonio Guterres points out, 'collective suicide'. If humanity is to tear itself away from rampant devastation of natural resources, it might start by changing its relations with animals to forge ecological adaptation and sustainability for its and the earth's continued existence.

If the extinction of animal species continues to rise, then the cross-pollination and germination of the world's forests and grasslands will lack diversification and slowly die out. How humans comprehend animals as intrinsic partners to inhabiting the earth is dependent on no longer seeing them as a source of food and as pets. Rethinking existing industrial farming of animals such as cattle, sheep, pigs, and fowl and reducing animal consumption is part of reducing human impact on the earth from methane gases in the atmosphere to deforestation for grazing pastures and non-native hoofed animal degradation of fragile soils and grasslands. There are also the health benefits humans would derive from vastly reduced consumption of animal products such as high blood pressure, heart disease, diabetes, and colon polyps. Besides these is humans' reappraisal of animals, to look on them not through the abyss of miscomprehension but to see them as our equal. Combining animal with human comprehension into a singular codependent existence can lead to a far greater awareness of the earth's ecology, informing new avenues for humans in how to live. Adjoining human relations with animals is integral to humanity's transition and transmutation in the 21st century. It is also dependent on redefining our partnership with the ecology of plants and how we might learn from their vast diversity to further derive a pathway for sustainable human existence.

Plant—Society in Botany

> *The plant-child, like unto the human kind—*
> *Sends forth its rising shoot that gathers limb*
> *To limb, itself repeating, recreating,*
> *In infinite variety; 'tis plain*
> *To see, each leaf elaborates the last—*[31]

In 1776, the then 26-year-old German philosopher, writer, poet, colour theorist, and botanist Johann Wolfgang von Goethe received an offer from the Grand Duke of Saxe-Weimar-Eisenach, Karl August to join his court. Aching to leave his stuffy city life in Frankfurt for small-town Weimar, Goethe happily accepted the offer. The duke provided Goethe with a home and importantly a garden, giving him the opportunity to pursue his interest in plants. Fourteen years later in 1790, Goethe published his findings from various botanical projects he undertook in his book *The Metamorphosis of Plants* that included a poem, which, as the selected lines mentioned earlier indicate, sought to elevate plant life as replete with human qualities.

The Metamorphosis of Plants is a meticulous recording of Goethe's experiments carried out in his garden. Goethe's botanical work on the metamorphosis of plants 'by which nature produces one part through another, creating a great variety of forms through the modification of a single organ'[32] would 70 years later find a link to Darwin's *On the Origin of Species* published in 1859. Darwin's transmutation of the species formulated his concept of natural selection where life forms were not static but evolved with the changing conditions of their habitat. Transmutation—the process of changing into another state or form—and metamorphosis—the process of changing by natural means from one form into another—might appear to be the same thing but they differ widely in terms of how and to what degree the environment conditions their transformation. Transmutation leads to the diversity of species, while metamorphosis leads to the diversity of variants. Where a seed can metamorphose to produce a new plant variant, the transmutation of an animal caused by the changes in its natural environment—hot, cold, arid, tropical, terrain characteristics, and food source availability to name a few—produces not a variant of the previous animal but an entirely new animal species. Goethe explains his metamorphosis concept through the sex of plants whereby 'nature steadfastly does its eternal work of propagating vegetation by two genders', whereas Darwin's Transmutation Theory ties species evolution to environmental conditions.[33]

Friend and peer of Goethe and predecessor to Darwin is the German biologist, botanist and geographer Alexander von Humboldt. Suffering from what he described as the stillness of European life and having been left a substantial amount of money following the death of his mother, 30-year-old Humboldt used his wealth to conceive a journey of scientific discovery. Outlining the programme, Humboldt wrote the following in a letter to his Berlin bankers:

> I will collect plants and animals, measure temperature, the elasticity, the magnetic and electrical content of the atmosphere, dissect them, determine geographical longitudes and latitudes, measure mountains. But this is not the main purpose of my journey. My real and sole purpose will be to investigate the interconnected and interweaving natural forces and see how the inanimate natural world exerts its influence on animals and plants.

Accompanied by the French botanist, Aimé Bonpland, and carrying 40 state-of-the-art instruments and collection jars, Humboldt and Bonpland departed from the port of A Coruña in Spain in 1799 to begin their five-year journey to South America. First sailing to the Canary Islands, they then crossed the Atlantic to land in Cumaná, Venezuela, six weeks later before sailing onward to Caracas. In Caracas, they prepared for their journey that would take them south to the Orinoco River, canoeing down this vast waterway to its outlet into the sea then overland back to Cumaná. Departing Cumaná they sailed north to Havana then south to Cartagena in Colombia, travelled overland through the Peruvian Andes to Lima then sailed to Acapulco, journeyed overland to Mexico City and Veracruz. Leaving Veracruz, they sailed to Havana then onto Washington before crossing the Atlantic back to Europe arriving at the port of Bordeaux in 1804. 'From my earliest days I felt the urge to travel to distant lands seldom visited by Europeans', Humboldt wrote in his travel biography. 'This urge characterizes a moment when our life seems to open before us like a limitless horizon in which nothing attracts us more than intense mental thrills and images of positive danger'.[34]

Stephen Jackson's introduction to Humboldt's and Bonpland's article 'Essay on the Geography of Plants', originally published in *The Future of Nature* in 1807, draws on Humboldt's ecological synthesis of nature where life on the earth is a unified system to humanity's impact on the environment.

> We now recognize the effects of atmospheric chemistry (including green-house gases such as carbon dioxide and methane), vegetation cover, and human land-use on climate at regional to global scales. We have learned that greenhouse gas concentrations are influenced at various time scales by tectonic activity, rock weathering, sequestration in soils and sediments, combustion of fossil fuels, absorption by ocean waters, fixation by vegetation, release by wildfires, and industrial synthesis of novel compounds (e.g., chlorofluorocarbons). We have discovered important feedbacks, negative and positive, among the various components of the system— including some anticipated by Humboldt, who speculated that effects of vegetation physiognomy on wind velocity, light reflection, and evaporation could affect climate at a regional scale.[35]

Whereas Goethe conceived his biological concept on the metamorphosis of plants from observations of his botanical interventions in seeds and plants in his garden in Weimar, Humboldt and Bonpland developed their ecological understanding of the lives of plants on a grand scale of field research in South America. Through their meticulous collection of plants and soils and recording of weather temperature, moisture, light, and wind, they were able to develop their conception that the earth is bound in a unified system of the elements and all living things. Humboldt and Bonpland explained their concept as follows:

> This is the science that concerns itself with plants in their local association in the various climates. This science, as vast as its object, paints with a broad brush the immense space occupied by plants, from the regions of perpetual snows to the bottom of the ocean, and into the very interior of the earth, where there subsist in obscure caves some cryptogams that are as little known as the insects feeding upon them.[36]

Collecting more than 1,500 samples though declaring this to be a drop in the ocean, Humboldt and Bonpland assumed the diversity of plants to be infinite like the cosmos. There was a growing interest on the part of Europeans across the sciences to chart the newly 'discovered' lands. Thirty years before, on the other side of the world, English botanist, Sir Joseph Banks' southern equatorial journey with Captain James Cook brought back a vast collection of plants and scientific advocacy to bring an awareness of the complexity and diversity of plant life to the scientific community of the time. Although Banks and Humboldt never met, the latter frequently met with Goethe to share their botanical knowledge of plant life and through their collective writings they elevated botanical sciences to become on a par with the disciplines of geology, cosmology, medicine, chemistry, and physics of the time.

Jumping forward to the 21st century, the Canadian environmental scientist Suzanne Simard published her book, *Finding the Mother Tree: Uncovering the Wisdom and Intelligence of the Forest*, which confirmed what Humboldt, Bonpland, Goethe, and Banks had believed concerning the society of plant life as an expansive interconnected entity. 'One of the first clues came while I was tapping into the messages that the trees were relaying back and forth through a cryptic underground network', Simard writes. 'When I followed the clandestine path of the conversations, I learned that this is pervasive through the *entire* forest floor, connecting all the trees in a constellation of tree hubs and fungal links'. Simard's discovery of an underground communication superhighway revealed what she called a 'crude map' where 'the biggest, oldest timbers are the sources of fungal connections to regenerating seedlings. Not only that, they connect to all neighbors, young and old, serving as the linchpins for a jungle of threads and synapses and nodes'.[37] Simard's thesis is relatively simple: the forest remains healthy by the diversity of tree species through its underground fungi networks sharing carbon, sugar, and nutrients and when threaten by infectious agents, are better able to fend off pathogens through sharing anti-viral protection. Simard was able to overturn the accepted ecological/botanical belief that competition was the supreme motivation for survival between plant species. It may not come as a surprise that trees share a quarter of their DNA with humans and the protection 'mother trees' provide to new saplings via their underground network rest ensures their survival and that of the forest. It is perhaps a humbling lesson for humanity to acknowledge the immense network of plant life intelligence covering the planet could be similar to theirs. The belief that all life on the earth is connected also harbours humanity's fears in the belief that its supremacy over all living things is all around them and as them. If forests communicate, share information, provide anti-viral protection, and protect their young, asks what is humanity doing to protect the next generation who will inhabit the earth? What the society of plant life tells us is if we don't protect the people who are struggling with the impact of climate change, inequality, and discrimination of all kinds, then humankind suffers.

To understand Simard's plant communication does not require a leap of faith to acknowledge that the society of plant life has obvious connections to human society—a porosity formerly denied. Similar to the experiences of Jane Goodall and Dian Fossey, it took Simard a great deal of effort to convince fellow environmental scientists and botanists concerning the different roles that fungi types play in the transmittance of carbon between trees. Simard's ground-breaking research confirmed what she had suspected, namely that plants relate to, depend on, and adapt throughout the seasons to shape their interconnected society. Her research tells us that felling trees is not simply about chopping them down, but also entails the eradication of diversity, communication, and society of plant life. What Simard's work makes clear is that the society of plants contains intelligent life forms commanding respect and protection, and that to decimate a forest is to decimate the 'intelligence' of the human intellect.

Integrating human society into the society of plants is one way to move from the decimation of forests and animals that has characterized human life on Earth. Humans, animals, insects, fish, trees, and plant life to oceans, geology, and geography are intertwined in a natural system where its preservation is imperative to human life. What should have been obvious but had escaped human comprehension has been brought to light in the face of the threat that humans and the earth now face. The environmental impact humans have raged across the earth's surface and how they have polluted its atmosphere reveals a lack in understanding of plant and animal intelligence in relation to their environments. Early forms of human intelligence with the natural world prevailed in early nomadic life and though it might not have had the botanical understanding uncovered by the likes of Humboldt, Bonpland, Goethe, or Simard, there would have been an understanding of a direct shared connectivity between them. To connect human intelligence today with the intelligence embedded in the natural world would require a massive change in how humans understand plant life. This would lead to a complete remodelling of how humans live on the earth, moving away from environmental degradation to associative co-habitation to lay the path for a collective sustainable future.

The metamorphosis of plants, creating new variants enabling their transformation for adaptation, does not belie their sedentary life; they are also mobile as seeds in the journeys of birds, bees, and animals in their continental and transatlantic migrations. In her introduction to *Moving Plants*, Line Marie Thorsen notes: 'Plants move, they move other things, they move people, and they are themselves being moved around'.[38] The physical mobility of plant life around the world is also realized in their sedentary yet mobile occupation of ground in the places where they grow. Planting a pine tree in the ground will no doubt grow in the same spot all its life, but its functions are wider. It will provide shade, protect and encourage other plants to flourish, adapt to changing weather conditions, and metamorphose as required, generating new shoots to forge an evolving ecological society below and above the ground. Simard and Thorsen are not alone in raising awareness of the mobility of plants especially concerning their lines of communication. Early scientific and philosophical records from ancient Persian and Greek writings reveal how scientists and botanists acknowledged the hidden society of plants as expressed by Aristotle's 'liminal soul'—the vegetative or nutritive soul that constitutes growth, nutrition, and reproduction reflecting human life. In *The Hidden Life of Trees: What They Feel, How They Communicate*, Peter Wohlleben describes the society of trees where the rate of photosynthesis is not the same for all trees but equalized between the 'strong and the weak', drawing on a kind of social security thread that connects them. 'Whether they are thick or thin, all members of the same species are using light to produce the same amount of sugar per leaf', he writes.

> This equalization is taking place underground through the roots... Whoever has an abundance of sugar hands some over; whoever is running short gets help. Once again, fungi are involved. Their enormous networks act as gigantic redistribution mechanisms. It's a bit like the way social security systems operate to ensure individual members of society don't fall too far behind.[39]

Humanity's blind perseverance in gaining superiority over nature has transpired to disregard plant life societies (and all living things) as remotely mirroring their own. Plants are the dominant living force of life on Earth as water is to oceans and terrain is to geography. As a part of the earth's diversity of life forms, humans rose through their capacity for innovation to dominate the natural world, shifting from coexistence to control over all living things. Shaping the earth to their design, humans irrevocably altered plant ecologies, disrupting the natural balance of the earth's biosphere to dramatically alter the weather we experience today under climate change. The global instability of weather patterns oscillating in the atmosphere due to carbon pollution is a direct result of humanity's rapacious instrumentalization of the natural environment to service its needs in the face of ecological collapse. To stop further environmental destruction, humans are deploying technological innovation to supervene their impact. A new belief system has emerged that is neither plant nor animal to re-shape human impact on the earth, that is, a belief in technologies such as carbon-capturing plants, digital programming of water management, genetic seed engineering, satellite weather monitoring associated with human programming of environmental control. Recalling Berger's animal comprehension of humans as man's abyss 'looking across ignorance and fear', that is, when *being seen* by the animal, he is being seen as his surroundings are seen by him', this can also be attributed to how humans look at plants through a non-apprehension of indifference.

In *The Life of Plants: A Metaphysics of Mixture*, Emanuele Coccia takes plants' lack of human features as betraying human indifference, something that comes at the expense of their fundamental importance to human life: 'They don't have hands with which to shape the world, yet it would be hard to find more capable agents when it comes to the construction of forms', Coccia tells us.

Plants are not only the most subtle artisans of our cosmos, they are also the species that have given life to the world of forms—they are the form of life that has made the world itself a site of infinite figuration. It is in and through plants that the earth has asserted itself as a cosmic laboratory, a space for the invention of forms and the making of matter.[40]

The human capacity to invent tools, design and produce products, conceive technologies, flatten topography to build cities has done much to rival the natural world or even to outdo it. The human desire to shape the earth and process its resources is built on overcoming any residual traces that characterized early humans' evolutionary inferiority to nature. 'Plants coincide with the forms they invent: all forms are, for them, inflections of being, and not merely of doing and acting', Coccia explains.[41] Humanity used its intellectual prowess to shape its surroundings as a way to manage plant, animal, and mineral resources. 'The plant is nothing if not a transducer, one that transforms the biological fact of the living being into an aesthetic problem and makes of these problems a question of life and death'.[42] Coccia's reference to the life of plants as the cosmos of the earth—the visible green seen from satellites—is but one half of the invisible underground network that makes up the other half.

When humans walked out of Africa approximately 100,000 years ago and traversed the world, separating and settling, they grew their knowledges of plant life in the environments they encountered. Plants were carried on human migration routes and the formation of different races, languages, culture, beliefs, and ceremonies was intricately connected to the diversity of plant and animal life as they were dedicated to the mysteries of the cosmos. Human intellect did not emerge on its own; it is the result of the intellect of the natural world and all living things. If humans had incorporated rather than dismissed the complex communicative network of plant life as equal to their own, there would be a far greater proportion of coexistence than the minority still held by remaining indigenous cultures around the world. From the earliest use of fire by humans 400,000 years ago to the establishment of settlement 12,000 years ago, humans set their evolutionary course through the belief that all things must be submitted to their control. As settlements grew into citadels and citadels into cities, the subversion of nature increased and the connection to its networks erased. Across the world, cities now absorb half of the world's eight billion people who, on a daily basis, have a limited experience of nature. Having conquered their creation in forging great human metropolises across the earth, humans turned to something greater than themselves to the expanses of the cosmos and planets as places of exploration and possible future habitation. The inferiority complex humans held towards the natural world, while still knowing so little of it, is now projected to intergalactic sites. Humanity's attempt to grasp its fragile existence has dominated its interaction with the natural world by altering its form, and in doing so, it has succeeded in managing its non-comprehension of the abyss of animal and plant life in relation to its own.

The society of plant life presents a universal language of integration and diversity through a shared supportive habitat spread across the world. The work of botany reveals the infinite and complex life of plants, yet humans see plant life as a resource for consumption or an arranged natural beauty to admire. We now understand that the devastation of plant life as with animals, land, and oceans will eventually devastate human life. Human design on the earth has continually dismissed the most omnipresent system of complexity of the earth: its plant life. New research in biotechnology and biomaterials is revealing the complexity of plant life systems to cultivate new material functions. Nature's inventory of life forms is diminishing and its ability to sustain networks across terrain is becoming increasingly uncertain. In less than 200 years, human impact has derailed the diversity and survival of plants' and animals' capacity for metamorphosis, transmutation, and adaptation—a capacity they have enjoyed since the formation of the earth 4.55 billion years ago. The instability of the natural environment in the 21st century belies in the unnatural turbulence oscillating in the atmosphere—both spelling the fragile future and end of human existence on the earth. Now faced with its potential demise, humanity is seeking to comprehend itself with the natural world; to seek the support of plant and animal life for its survival even while it keeps on destroying it.

Plant and animal societies offer humans solutions to deal with the excesses of their consumption of the earth, habitation, and co-communicative coexistence. Animal and plant life's capacity for transmutation, metamorphosis, and adaptation to changing environments offers pathways to stop the freefall of environmental devastation. Given that humans are not genetically capable of transmutation or metamorphosis but are capable of undergoing transformative adaptation, this will, if undertaken, limit further climate change damage. But in order for humans to incorporate plant and animal complexity into their lives, there is a need to learn how to resist continuous growth and consumption. To help remedy human impact on the environment, people need to view themselves as extensions of animal and plant ecologies and the stability of weather that binds them together in an interconnected network— the kind that Humboldt, Bonpland, Goethe, indigenous peoples across the world, ancient Greeks, Mesopotamians, and many other cultures knew to be the core to achieving global human–environment sustainability. To do this, humans must give up the central position they have occupied in their relations to the natural world and grasp their own comprehension as animal, plant, and weather.

Human—Animal-Plant-Weather

Since the formation of Earth 4.55 billion years ago, its land mass has stretched, contracted, and separated across the surface bridging oceans, seas, and islands to form the diversity of climatic conditions north and south of the equator. During human migration across the earth from Africa 100,000 years ago to the ending of the ice age 11,500 years ago, almost every part of the earth with the exception of Antarctica, the North Pole, and some isolated islands has been populated by humans. Human race, culture, and language diversity were forged by a combination of weather, geography, oceans, food, ceremonies, and the cosmos. Human survival was based on gathering available animal and plant resources specific to the region for food, clothing, and protection against the weather. Weather shaped early human life and all living things including sediment rock by their exposure to it. From the making of shelter to the building of cities, humans learned to build-out the weather and build-in climatic environments to temper the effects of hot, cold, wet, and windy conditions. As animal and plant life thrived and died in extreme weather events, humans could seek relief in the relative comfort of their shelters. However, not even the vast sealed spaces of cities that later emerged could shield them from weather turbulence caused by 200 years of rampant industrialization. In this extremely short period of unparalleled progression, human existence became suddenly precarious due to its negative impact on the world's ecology.

The first signs of humans' ability to control their relation to weather was to shield their exposure from it. From hollowing out the earth to crafting from brush, wood, and stone, humans have adopted behavioural traits to manage weather. Animals and plants on the other hand either adapted and flourished or perished by their exposure. Weather governs every aspect of the earth, from the life of all living things to the geology of stone and geography of deserts, glaciers, tropical forests, and tundra. Weather is the accumulative power of the elements and everything on the earth submits to its forces. Early human existence relied on its relationship to weather, animals, and plants for survival. Animals provided sustenance and by skinning their hide, using animal skin to cover human skin, humans outwardly appeared animal. Wearing animal, humans cloaked themselves to the weather giving them greater protection to seasonal changes, and in appearing animal, humans mirrored the animals they hunted. By wearing animal hide, humans wore the ecology of the animal and its environment. The distinction of the 'non-comprehensive of the abyss' separating humans and animals had not yet taken hold in early human existence. Over time, wearing animal hide became more complex in adaptive patterning to the weather; it could be cut and sown to the contours of the body. Throughout human migration, humans adapted and incorporated the animal into their own, and since the establishment of settlement 12,000 years ago, the cultivation of crops and the crafting of textiles, humans slowly wore-out their animal skin to the folds of fabric and the intricacy of seams. Once animal-like and exposed to the elements, humans separated from the wild connective habitat of animals and the weather. In a short period of time in human evolution, the instinctual coexistence between humans, animals, and the weather was no longer determined by their shared ecological skins.

Climate change has taught humans that whatever clothing they wear and regardless of whether they reside in the sealed compartment of their shelter, weather is inescapable. Nevertheless, humans have successfully negotiated the weather by constructing their own artificial meteorological environments to immobilize it at the barriers of concrete and glass of their homes, work environments, compartments of transport, shopping malls, stadiums, and indoor swimming pools that operate in all weather conditions. This conditioning of weather is not a blanket experience for weather exposure is an inherent part of the rural life of farmers and herders, the dispossessed, homeless, refugees, and first nations peoples around the world. In today's global industrial consumer-led economy, the elemental forces of weather are measured in degrees of productive assistance and profitability or conversely disruption and devastation. In rural areas, weather directly provides the means of survival, providing the right conditions to grow crops or unleashing floods and droughts to destroy them.[43] Weather aids and abets, supports, and threatens human survival.

Forecasting weather through data collection from satellites orbiting the planet has allowed for weather to be mapped and patterned to better inform human life on the earth. Televised weather forecasts have brought the science of meteorology in real-time analysis of highs and lows in atmospheric pressure, currents, ocean, and land temperatures to reach the homes and smart phones of billions of people. In *Weather as Medium: Toward a Metrological Art*, Janine Randerson writes: 'Weather exists in the media in the familiar context of *weather as news*, in which the forecasting of weather through maps and broadcasts structures our daily activities'.[44] Meteorology—the science of weather—and metrology—the science of measurement and weight—are combined in the study of the atmosphere. In *Meteorology Today*, meteorologist Donald Ahrens notes: 'We live at the bottom of the troposphere which is an atmospheric layer where the air temperature normally decreases with height. The troposphere is a region that contains all of the weather we are familiar with'. 'When we talk about the weather, we are talking about the condition of the atmosphere at any particular time and place. Weather—which is always changing—is comprised of the elements of: air temperature, air pressure, humidity, clouds, precipitation, visibility and wind'.[45] Displayed via charts of swirling blue, red, ochre, and green signalling cloudburst, torrential rain, hurricanes, tornados, extreme heat, and dry conditions, the daily weather forecast as news beams into the households of billions of people the natural movements of the earth's elements conditioning the planet at any given time.

Weather reporting and climate forecasting work on two distinct timeframes in short- and long-term predictions. The information each imparts lies between the tangible day-to-day experience of weather and accumulative statistical modelling to predict patterns. Reports of climate change effects sit in both the short- and long-term predictive analysis but not in the day-to-day timeframe. In her book *The Future of Weather*, climatologist Heidi Cullen points to the problem of how the public can understand climate change when it is absent from the daily weather reporting.

> Climate and global warming need to be built into our daily weather forecasts because by connecting climate and weather we can begin to work on our long-term memory and relate it to what's outside our window today. If climate is impersonal statistics, weather is personal experience. We need to reconnect them.[46]

Images of spiralling cloud formations, hurricanes, tropical cyclones, extreme heatwaves, cloudless skies bearing no precipitation have become ever-more common as a result of climate change. In one part of the world, flash floods wash away houses, crops, animals, and plants and in another part of the world dry earth, decimated crops, dead animals, and emaciated human bodies starving from lack of food fill the televised reports of an increasingly unpredictable global weather system. Catastrophic forecasts in some part of the world and the consequences predicted to befall a population living in vulnerable regions are relayed around the world through global communications. The frequency of disastrous weather events and the gap between victims and the protected displays the stark realities between imagery and experience, the attitudes of despondency and indifference when it comes to the effects of climate change. Weather is being divided between the filmic in the charts and images of televised projections as against the devastating realities of lives affected. There are, of course, benefits of mediatized weather connecting humans to a global knowledge of the earth relayed as sets of graphics. In response to the catastrophic effects of climate change, weather forecasting has become commoditized into streams of turbulence of apathetic dissociation as much as warnings to make timely preparations ahead of weather events.

Physical and yet ephemeral, powerful, and passive, the capricious nature of weather has always been naturally volatile and destructive. What has become clear now is that its unnatural and unpredictable volatility on a wholly unstable Earth due to climate change has shaken humanity. Weather now threatens on a global scale the lives of people, regions, countries, and continents. Once snow-capped and glacial mountains now reveal only rock, rivers run dry, and encroaching desert sands cover once fertile land. In acknowledging that weather as the defining elemental force never rescinded from its central place in creating and ceasing the life of all living things on the earth, it does so now with extreme velocity towards destruction. Fluctuations in weather have always negatively and positively affected the surface of the earth and provided an impetus for animal and plant life species diversity through environmental adaptation and metamorphosis. But the effects of climate change are disrupting the evolutionary cycle of species. Now more than ever, weather is turning regions once inhabitable into inhospitable places in increasingly shorter time periods. Human-made toxicity of the atmosphere is now patterning humanity's existence—this is human-made destruction.

Figure 4.1 Hurricane Ian, cloud formation, Caribbean Sea east of Belize, 26 September 2022.

Source: Image courtesy of NASA.

Weather is the interface through which humans experience the earth. It is the bridge between the physical realm of geography and the felt, temporal, and ephemeral velocity of the elements. Inescapable, weather has become an immaterial invention of the 'pictorial', able to be seen but not felt behind plates of glass of seamless interiors, passed through via highways in cars, and air-lifted into the atmosphere in pressurized capsules of aircraft; each forging an indifferent relation to the outside world. While turning weather into pictorial representation and numerical measurements is invaluable to farmers, fishermen, shipping companies, and the like, it is devalued in many people's lives especially those living in cities. Possible predictive futures of degraded landscapes, melted glaciers, dry rivers and lakes, disappearing islands, and coastal cities under rising oceans is not the sphere of mediatized weather. Unpredictable, volatile, and destructive, weather must now be used to bring city and rural populations together rather than continue their current divisions of expulsion and exposure. In cities where weather is filtered so as to rarely alter or threaten human life, weather in rural regions infiltrates and patterns the lives of the farmer, grazer, and herder, aiding or abetting their ability to provide sustenance for survival.

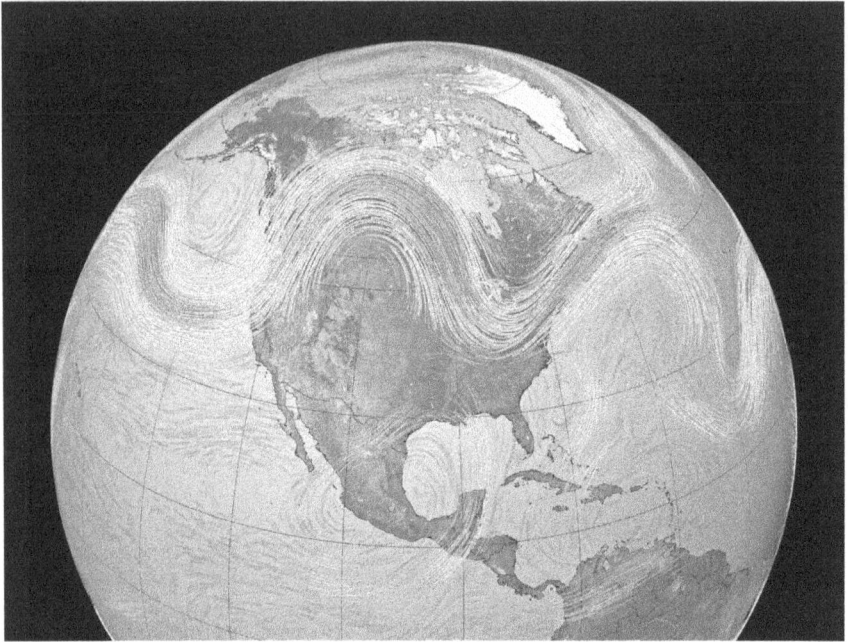

Figure 4.2 Polar jet stream global map. Trent L. Schindler, 3 November 2011.

Source: Image courtesy of NASA.

The first signs of climatic impact on the earth were evident in extreme changes in the weather. As weather turbulence increased, so too did its ability to threaten human, animal, and plant life. To pull humanity back from the brink of environmental cataclysm, it is obvious we must harm less—if at all—the natural ecosystems of the earth by reconciling with weather and thereby reconciling with the earth. Reconnecting with animals, plant life, and weather is one way out of the abyss of non-comprehension that humans have built in their trajectory for dissimilation to the natural world of which they were once firmly and intrinsically a part. Equally cultivators and destroyers, humans have still much to learn about how our existence on the earth can be sustainable to secure the future of human, animal, and plant life. A part of learning from animals and regaining 'our' animality, humans must learn from the societies of animal and plant life that have existed, adapted, and diversified to their surroundings. Humans must also relearn how to assimilate with the natural world, to walk the ground, to collect geography as akin to gaining Earth knowledge, to migrate and adapt with animal and plant life, and to experience weather as elemental to human survival.

Alexander von Humboldt understood through observation, analysis, and botanical collection that the earth is a unified system, just as Goethe understood that the metamorphosis of plants creates the diversity of new life, and Suzanne Simard broke new ground concerning the complexity of plant life interfamilial societies. The works of each of these botanists and ecologists contain messages to guide and transform humanity's impact on the earth. Berger and Shepard pointed to our non-comprehension of animal from the abyss as much as a pathway in finding new directions for the ecology of human and animal assimilation in the natural environment. Abrams, Massumi, and Bourke noted the animality of the human to portray indifference as projected security but which is in fact the continuation of the fear and timidity that has followed human existence throughout its evolutionary path. Weather has suffered the fate of humanity's indifference through humans' ability to dissect and disassociate the elements through building—dividing experience between the material surfaces of inside and outside. Weather, animals, and plants; how we look at, understand, and experience the natural environment and the oscillating atmospheric conditions that pattern life on Earth—to acknowledge all of these is to recognize that the separations between us are no longer sustainable.

The opportunity to become animal and plant by associating with their societies at a deeper level is one way in which humanity can shift away from its central position on the earth and towards a more peripheral place. For human and Earth co-existence to endure, weather has to be worn, and urban and rural populations connected to move across the same surface of the earth without separation. Through a process of collective osmosis, humans can rework themselves into animal and plant life as equal 'citizens' of the earth. To start this journey, humanity must reject its 'nature' of rampant consumption, assume new economic and social models, and stabilize the earth's weather patterns so that a future sustainable world may emerge. The following chapter, 'Climate Gathering—*wearing our ecology*', works in and out of the structural divisions of human and environmental displacement through the cartographical divisions of terrain and ground mobility to the concept of environmental costuming to suggest how humans *become* their ecology.

Notes

1 Ernst Haeckel, *The History of Creation, or, the Development of Earth and Its Inhabitants by the Action of Natural Causes*, Vol. 1, translated by E. Ray Lankester. London: Henry King & Co., 1876, pp. 44–45.

2 *Quran*, 6:99 Surah, Chapter 6, Al-An'am, verse 99 Juz, part 7, p. https://quran.com/6/99

3 Jeffrey Jerome Cohen. *Stone: An Ecology of the Inhuman*. Minneapolis: University of Minnesota Press, 2015, pp. 14–15.

4 Ibid., pp. 20 and 21.

5 Thomas Nail. *Theory of the Earth*. Stanford: Stanford University Press, 2021, p. 177.

6 Ibid., p. 178.

7 John Berger. *About Looking*. New York: Pantheon Books, 1980, p. 1.

8 Ibid., pp. 3 and 5.

9 Paul Shepard. *Thinking Animals: Animals and the Development of Human Intelligence*. New York: Viking Press, 1978, p. 32.

10 Shepard points to the society of animals as to the society of humans in a combination of language and behaviour.

> Through language, common terms appear for relatedness (sex and nutrition), history (genealogy and the animal life cycle), membership (clan and species), individuality (personality and species behavior), role (gender and age and niche), drawing simultaneously from within the society and from the ecology without. Each of these is incomplete without the other.
>
> (Ibid., p. 37)

11 Shepard outlines humankind's cruelty towards each other as a precursor to their cruelty towards animals.

> Ethically, there have been discourses against the enslavement, torture, and killing of people since civilization began without ending war, tyranny, or cruelty. There is no evidence that crime, brutality, or murder have diminished at all. If human behavior is not improved by the incorporation of such ethics into the dominant religions, what reason is there to suppose that such a new ethic can save animals? The very ideology that raises the importance of every individual and seeks the nobility in our species can be used to support the need to exploit animals, if need be, for the benefit of the most noble species. The avoidance of unnecessary cruelty, insofar as that is done, will make the moralists feel better and reduce pain, but it will not save animals.
>
> (Ibid., pp. 237 and 248)

12 Ibid., p. 261.

13 See Paul Shepard. *The Others: How Animals Made Us Human*. Washington: Island Press, 1997, p. 90.

14 Ibid., pp. 53–54.

15 Ibid., p. 250.

16 Ernst Mayr. *Animal Species and Evolution*. Cambridge, MA: The Belknap Press, 1963, p. 5.

17 See Laurence Simmons' chapter 'Shame, Levinas's Dog, Derrida's Cat (and Some Fish)'. In *Knowing Animals*, edited by Laurence Simmons and Philip Armstrong. Leiden: Brill, 2007, p. 37.

18 See Ricardo De Vos' chapter 'Extinction Stories: Performing Absence(s)'. In *Knowing Animals*, edited by Simmons and Armstrong, p. 183.

19 Ibid., p. 184.

20 Besides the introduction of cane toads to Australia there was also the 'cactoblastis moth to fight the prickly pear (a noxious introduced weed species) and the introduction of the myxoma virus, which effectively reduced the number of feral rabbits in the 1950s'. All of these were to have devastating effects on the native

ecological system. See Catharina Landström. 'Australia Imagined in Biological Control'. In *Knowing Animals*, edited by Simmons and Armstrong, p. 200.

21 See Sabrina Tonutti. 'Anthropocentrism and the Definition of "Culture" as a Marker of the Human/Animal Divide'. In *Anthropocentrism: Humans, Animals, Environments*, edited by Rob Boddice. Leiden: Brill, 2011, pp. 185–186.

22 Joanna Bourke. *What It Means To Be Human: Reflections from 1791 to the Present*. Berkeley, CA: Counterpoint, 2011, p. 10.

23 Ibid., p. 10.

24 Further to Bourke's Möbius strip analogy, to place human and animal on the same dialectical pathway turns to the historical argument of simulation under the disruptive evolutionary Theory of Descent developed by Darwin.

> Darwinian arguments may have contributed to the deconstruction of the radical differences imagined between humans and animals, but humanism survived this attack. It did this, in part, by rejecting absolutist narratives of the human (the claim that people are utterly distinct from animals) and embracing relativistic ones (the idea of a continuum between the two states, with the fully human at one end and the fully animal at the other).
>
> (Ibid., pp. 11 and 380)

25 David Abram. *Becoming Animal: An Earthy Cosmology*. New York: Patheon Books, 2010, p. 69.

26 Ibid., p. 93.

27 Further to Abram's conception of humans themselves as 'apart from the world', he also questions the truth of human knowledge to themselves and their trust in technologies.

> We've taken our primary truths from technologies that hold the world at a distance. Such tools can be mighty useful, and beneficial as well, as long as the insights that they yield are carried carefully back to the lived world, and placed in service to the more-than-human matrix of corporeal encounter and experience. But technology can also, and easily, be used as a way to avoid direct encounter, as a shield—etched with lines of code or cryptic jargon—to ward off whatever frightens, as a synthetic heaven or haven in which to hide out from the distressing ambiguity of the real.
>
> (Ibid., pp. 94 and 7–8)

28 Brian Massumi. *What Animals Teach Us About Politics*. Durham, NC: Duke University Press, 2014, p. 54.

29 Further to Massumi's reflective indifference between animal and human, he highlights the following where

> the zone of identificatory indifference serves as a medium for conveying a sameness of form. In the zoo visit, the anthro-form anamorphoses onto the animal. In the imitation, the movement goes in the opposite direction. It is the form of the observed animal that anamorphoses onto the human viewer, wallpapering it with an animal motif. This is a secondary reprojection—a distorted retrojection conditioned by a prior ana-anthropomorphizing projection. Only humans imitate animals.
>
> (Ibid., p. 82)

30 Expanding on Abram's evolutionary trajectory in reference to Darwin's animality transmutation regarding human evolution, he writes:

> Darwin had rediscovered the deep truth of *totemism*—the animistic assumption, common to countless indigenous cultures but long banished from polite society, that human beings are closely kindred to other creatures, and indeed have various other animals as our direct ancestors. Here was a form of totemism transposed into the modern world—the totemic insight now translated into the language of 'descent by natural selection from a common ancestor.' This modern version no longer saw different persons as descendants of *different* totemic animals, but recognized all humankind as derived from a common lineage of creatures. In the wake of Darwin's bold insights, we have learned to consider all humans as members of a common family. But the wild, animistic implication of Darwin's insight has taken much longer to surface in our collective awareness, no doubt because it greatly threatens our cherished belief in human transcendence.
>
> (Abram, *Becoming Animal*, p. 77)

31 Johann Wolfgang von Goethe. *The Metamorphosis of Plants*. Cambridge, MA: MIT Press, 2009, p. 2.
32 Ibid., p. 5.
33 Ibid., p. 65.
34 Alexander von Humboldt. *Personal Narrative of a Journey to the Equinoctial Regions of the New Continent*, translated by Jason Wilson. London: Penguin Classics, 1995, p. 15.
35 Alexander von Humboldt and Aimé Bonpland. *Essay on the Geography of Plants*, edited by Stephen T. Jackson, translated by Sylvie Romanowski. Chicago: University of Chicago Press, 2009, p. 42.
36 Ibid., p. 64.
37 Suzanne Simard. *Finding the Mother Tree: Uncovering the Wisdom and Intelligence of the Forest*. London: Penguin Books, 2021, p. 5.
38 Line Marie Thorsen ed. *Moving Plants*. Rønnebæksholm Næstved: Narayana Press, 2017, p. 11.
39 Peter Wohlleben. *The Hidden Life of Trees: What They Feel, How They Communicate*, translated by Jane Billinghurst. Vancouver: Greystone Books, 2015, p. 16.
40 Emanuele Coccia. *The Life of Plants: A Metaphysics of Mixture*, translated by Dylan J. Montanari. Cambridge: Polity Press, 2019, p. 12.
41 Coccia divests humanity's inventory of creativity, saying:

> Unlike higher animals, wherein development stops once the individual has reached his or her sexual maturity, plants never cease to develop and grow, to construct new organs and new parts of their own body (leaves, flowers, parts of the trunk, etc.), which they previously lacked or had gotten rid of. Their body is a morphogenetic industry that knows no interruption.
>
> (Ibid., p. 13)

42 Ibid.
43 Famine is based on the following criteria: acute malnutrition affecting more than 30% of the population, 20% households facing severe food shortages, and 2

or more deaths per 10,000 people per day. *Source*: https://guide-humanitarian-law.org/content/article/3/famine-1/. Crop failure due to drought, extreme heat, water scarcity is not the sole cause of famine but also human conflict where fighting displaces people from their lands leading to food insecurity. The causes of famine are the lack of rainfall, poor or inappropriate seeds, or degraded soil due to over-clearing of land reducing ground water retention. A severe drought is calculated when there is more than a 50% drop in rainfall over a season or year. Ongoing droughts like in the Horn of Africa, now exceeding five years, have caused the displacement and extreme food insecurity of tens of millions of people. For more concise information on drought declaration, see World Health Organization: www.who.int/health-topics/drought

44 Janine Randerson. *Weather as Medium: Toward a Metrological Art*. Cambridge, MA: MIT Press, 2018, p. xvii.
45 C. Donald Ahrens. *Meteorology Today: An Introduction to Weather, Climate, and the Environment*. Belmont, CA: Brooks/Cole, 2009, p. 18.
46 Heidi Cullen. *The Weather of the Future: Heat Waves, Extreme Storms, and Other Scenes from a Climate-Changed Planet*. New York: HarperCollins, 2010, p. 34.

Bibliography

Abram, David. *Becoming Animal: An Earthy Cosmology*. New York: Patheon Books, 2010.
Ahrens, Donald C. *Meteorology Today: An Introduction to Weather, Climate, and the Environment*. Belmont, CA: Brooks/Cole, 2009.
Berger, John. *About Looking*. New York: Pantheon Books, 1980.
Bourke, Joanna. *What It Means To Be Human: Reflections from 1791 to the Present*. Berkeley, CA: Counterpoint, 2011.
Coccia, Emanuele. *The Life of Plants: A Metaphysics of Mixture*. Translated by Dylan J. Montanari. Cambridge: Polity Press, 2019.
Cohen, Jeffrey Jerome. *Stone: An Ecology of the Inhuman*. Minneapolis, MN: University of Minnesota Press, 2015.
Cullen, Heidi. *The Weather of the Future: Heat Waves, Extreme Storms, and Other Scenes from a Climate-Changed Planet*. New York: HarperCollins, 2010.
De Vos, Ricardo. 'Extinction Stories: Performing Absence(s)'. In *Knowing Animals*, edited by Laurence Simmons and Philip Armstrong. Leiden: Brill, 2007, pp. 183–195.
Goethe, Johann Wolfgang von. *The Metamorphosis of Plants*. Cambridge, MA: MIT Press, 2009.
Haeckel, Ernst. *The History of Creation, or, the Development of Earth and Its Inhabitants by the Action of Natural Causes*, Vol. 1. Translated by E. Ray Lankester. London: Henry King & Co., 1876.
Humboldt, Alexander von. *Personal Narrative of a Journey to the Equinoctial Regions of the New Continent*. Translated by Jason Wilson. London: Penguin Classics, 1995.
Humboldt, Alexander von, and Aimé, Bonpland. *Essay on the Geography of Plants*, edited by Stephen T. Jackson, translated by Sylvie Romanowski. Chicago, IL: University of Chicago Press, 2009.
Landström, Catharina. 'Australia Imagined in Biological Control'. In *Knowing Animals*, ed. Laurence Simmons and Philip Armstrong. Leiden: Brill, 2007, pp. 196–212.

Massumi, Brian. *What Animals Teach Us About Politics*. Durham, NC: Duke University Press, 2014.

Mayr, Ernst. *Animal Species and Evolution*. Cambridge, MA: The Belknap Press, 1963.

Nail, Thomas. *Theory of the Earth*. Stanford: Stanford University Press, 2021.

Randerson, Janine. *Weather as Medium: Toward a Metrological Art*. Cambridge, MA: MIT Press, 2018.

Shepard, Paul. *Thinking Animals: Animals and the Development of Human Intelligence*. New York: The Viking Press, 1978.

Shepard, Paul. *The Others: How Animals Made Us Human*. Washington, DC: Island Press, 1997.

Simard, Suzanne. *Finding the Mother Tree: Uncovering the Wisdom and Intelligence of the Forest*. London: Penguin Books, 2021.

Simmons, Laurence. 'Shame, Levinas's Dog, Derrida's Cat (and Some Fish)'. In *Knowing Animals*, edited by Laurence Simmons and Philip Armstrong. Leiden: Brill, 2007, pp. 27–42.

Thorsen, Line Marie, ed. *Moving Plants*. Rønnebæksholm Næstved: Narayana Press, 2017.

Tonutti, Sabrina. 'Anthropocentrism and the Definition of "Culture" as a Marker of the Human/Animal Divide'. In *Anthropocentrism: Humans, Animals, Environments*, edited by Rob Boddice. Leiden: Brill, 2011, pp. 183–199.

Wohlleben, Peter. *The Hidden Life of Trees: What They Feel, How They Communicate*. Translated by Jane Billinghurst. Vancouver: Greystone Books, 2015.

5   Climate Gathering
Wearing Our Ecology

Border Lines—Climate Cartographies

The previous chapter explored how humans might benefit by shifting from their present comprehension of animals, plants, and the weather from separable and exclusionary associations to inseparable and inclusive coexistence. The chapter outlined that with the establishment of settlement and agriculture, humans shaped the environment to their needs, expediting their separation from animal and plant life and instituting their position as the superior entity on the planet. Spurred by European exploration and colonialism in the 16th and 17th centuries and consolidated during the Industrial and later Technological Revolutions from the 18th through to the 20th centuries, humans forged global inequality, unrestricted resource extraction, environmental destruction and cultural decimation, and enslavement of native peoples while processing resources into objects, building cities to house mega populations, and collapsing time and distance through modern transport and satellite communications. The impact of these trajectories has ensured that there are no places on the planet that humans have not touched and depreciated by altering, extracting, processing, and consuming its resources. This has brought on atmospheric pollution, land acidification and desertification, fresh water and sea contamination, among others. To counteract these histories, the previous chapter suggested how humans might reimagine their relationship with the natural world that characterized early nomadic life and in doing so avert further collapse of the earth's biosphere. This return to rebalance the negative human impact on the earth is predicated on the rediscovery of animal and plant societies as a new epoch in human evolution in the 21st century. This chapter explores how humans might incorporate the natural world as a part of us 'wearing our ecology'. Beginning with Border Lines—Climate Cartographies, the section seeks to redefine cartography of the map to the map of the body to present a new experience of terrain for a lived sustainability. The second section, Climate Moves—Migrating Sustainability, explores how the removal of natural surfaces have come to dominate human and ground separation and how they might be reinstated. The third section, Mobile Ecologies—Environmental Costuming, combines

DOI: 10.4324/9781003382515-6

cartography and movement to create an intrinsic geography of the body to formulate new modes of human inhabitation on the earth.

The 'Global South' emerged from the histories of European colonialism where continents, cultures, and people were divided and demarcated by cartographic lines, not on tribal lands or the natural characteristics of geography, but on the imposed formation of countries mostly notably in Africa, North and South America, Asia, and Australia. European expansionism turned tribal lands into sites for resource theft, dispossessed native peoples of their homelands, and submitted them to enslavement. The modern-day exodus of millions of people from their lands due to drought, water scarcity, and famine is part and parcel of the continuation of this exploitation and the writing of cartographical lines of separation. The central role that cartography has played, specifically on the African continent, is no more evident than in the 1884–1885 'carving-up of Africa' at the Berlin Conference and later Belgium Conference of 1890 by European colonial rulers.

European mapping of Africa separated the continent into spatial territories through a combination of errors in representing terrain and deceit in misrepresenting its original owners. The errors came by means of restraining geography to lines and marks to denote physical characteristics and deceit by restraining terrain to stasis against the movement of cultural exchanges over land between people. Cartography's representation of terrain limited geography to factual record—topographical lines, undulations, and the scaled measurement of distance to illustrate a 'groundless' representation to the physical reality of the ground. 'Maps present only one version of the earth's surface', James Corner writes in *The Agency of Mapping*, 'an eidetic fiction constructed from factual observation'.[1] While the speculative measurement of cartography reveals geographies, spatial composition, and dimensions, it does not show a biomorphic vision to evoke the living reality of the terrain and the living things within it. Spatial theorist, Edward Soja, suggests that spatial representations of ground such as maps are 'another form of environmental or geographical determinism'. He asks: 'How can fixed and dead urban geographies shape the dynamic development of social processes social consciousness, social will?'.[2] The process of cartography involves writing the ground to convey via lines and markings the lay of the land. In looking at the African continent, the map of its 54 countries born from agreements between imperial powers in the late 19th century has had profound ramifications in relation to climate change in the 21st century. Under colonial powers, land dispossession, forced labour in plantations, unlimited access to mining, and mass rural migration to urban centres has cut across cultures and traditions, with increasing vulnerability, exposure, and inability to adapt to severe weather events. The cartographic lines drawn over the African continent from agreements made in the 19th century run counter to a continent bearing the non-bounded effects of climate turbulence oscillating in the atmosphere. The non-territorial impact of global warming is now more than ever challenging Africa's cartographic map not only through the origins of error and deceit in

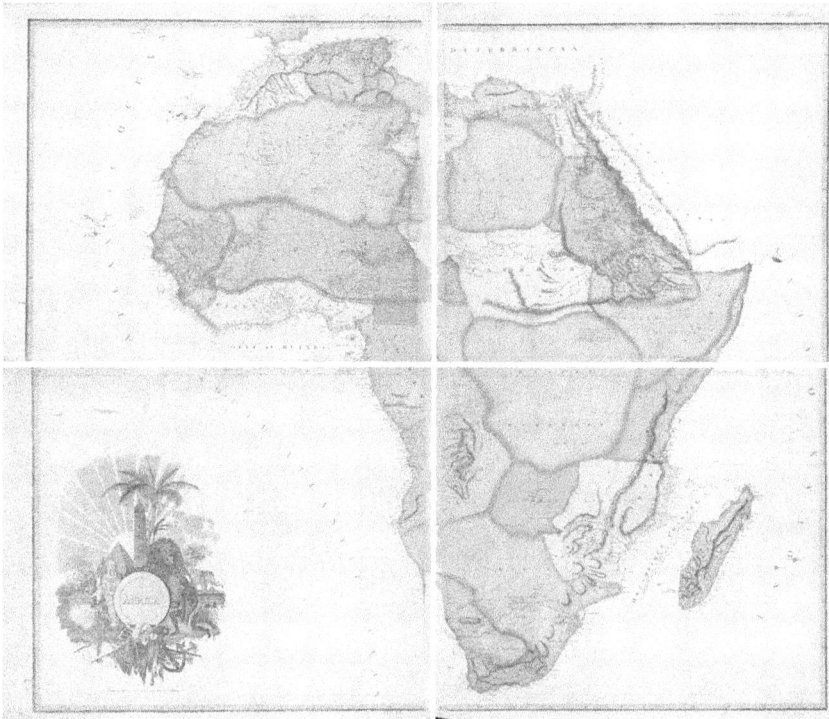

Figure 5.1 Africa, 1831. African continent divided by geographical regions. Map created by Henry Schenck Tanner.

Source: Image courtesy of Norman B. Leventhal Map & Education Center, Boston Library.

artificially constructing countries but also as an outmoded representation of the continent in terms of climate change.

Cartography provided humans with an 'off-the-ground' view of the world in plan. Contours and markings of topographical characteristics traced the route of rivers and lakes, mountains and plains, oceans and seas. It abetted the sciences of geology, geography, botany, and cosmology and constituted a powerful companion and weapon of imperialism to demarcate enforced territorial claims perpetrated on foreign ground from previously non-existing boundaries. Maps drawn in the 18th and 19th centuries sliced and ruptured first nations peoples' homelands throughout the world, severing the embodied connectivity to their environments, the remnants of which continue today. The American geographer, Bernard Nietschmann, referred to the power of cartography as follows: 'More Indigenous territory has been claimed by maps than by guns. And more Indigenous territory can be reclaimed and defended by maps than by guns'.[3] The effects of global warming are challenging the

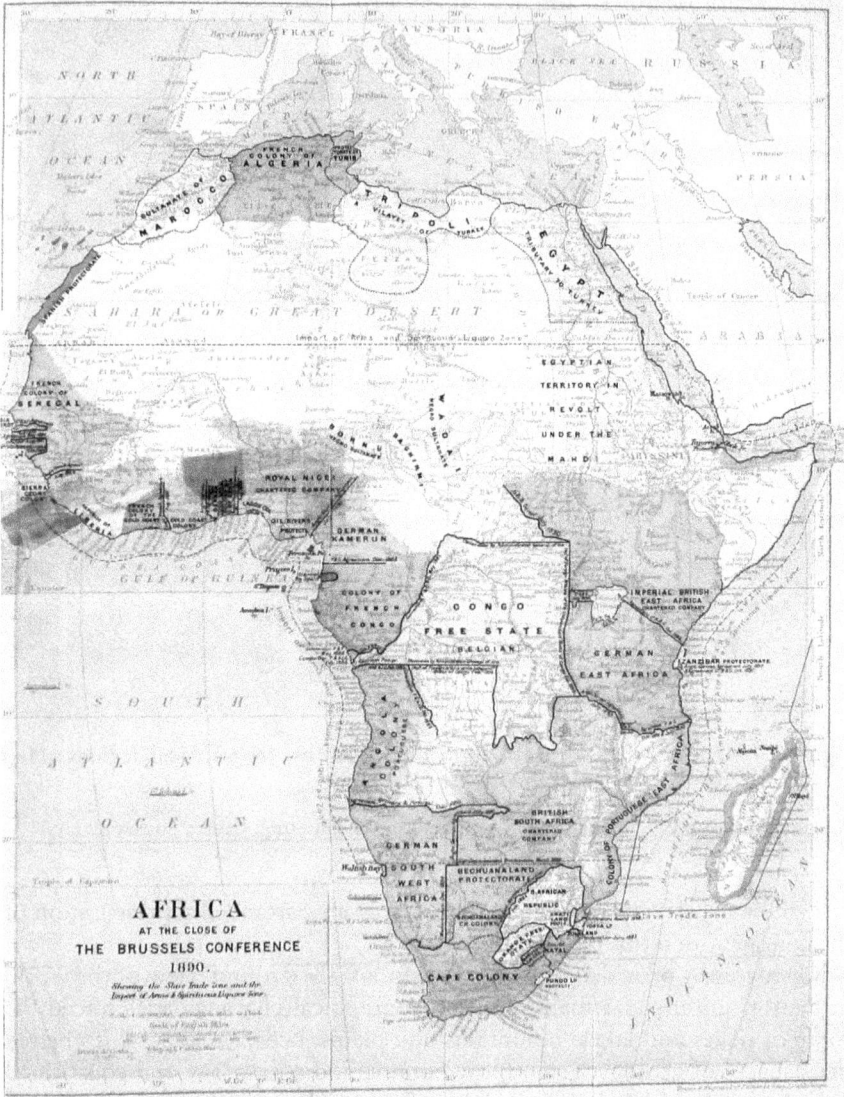

Figure 5.2 Partition of Africa, 1895 by Sir John Scott Keltie as a result of the 1890
   Belgium Conference.

Source: Image courtesy of Norman B. Leventhal Map & Education Center, Boston Library.

long-demarcated lines of maps and forging a new cartography of the world
drawn not by borders and boundaries defining countries but by regional
and geographical characteristics. This new cartographic representation of
the world would define the movements of climate change, giving territories

elastic boundaries and the people who occupy the land the ability to instigate new methods to avert further degradation and to implement sustainable adaptation.

If we were to draw a new map of the world today informed by climate change, it would show the earth divided into two parts, necessitating the drawing of two maps. One map would show regions of the earth relatively environmentally secure and ringed by protectionist borders, these being the countries of rich Western nations. The second map would show porous, fragile, and increasingly threatened regions, notably poor and developing countries, most vulnerable to climate change. The two maps would illustrate the divisions between the static and secure regions and their populations as against the shifting and fragility of regions and their populations affected by climate change, rewriting the world into zones of climate change causes and effects of the under- and over-exposed relative to their global position on the earth. It would become clear that countries least affected by climate change possess a greater capacity to financially provide support for climate change adaptation, build infrastructure, and instigate alternative energy production to those countries most impacted. The present global situation is something vastly different from what these new maps illustrate where under-resourced and ill-prepared poor and developing countries are most vulnerable to global warming, while the rich and developed nations in-build their climate change security whilst maintaining their standard of living, consumption, and by default gross output of global CO_2 emissions. Joining the maps together would show a cartography of the world where the border lines between countries do not correspond to the borderless and unpredictable patterns of weather turbulence on the ground. Such maps are in circulation and the work of geographers and cartographers at the pro-commons website, Decolonial Atlas, has created a vast archive of cartograms covering political, economic, social, cultural, and environmental issues as can be seen in this volume.[4]

Combating the effects of climate change through direct financing of regions removes overt national and international government interference and corruption. A way to finance at-risk regions around the world could be drawn from the cartographical map that identifies less affected and financially secure countries where a percentage of their overall GDP is funnelled to subsidize lands and people in the most threatened regions. The prospect of directly funding such regions is a far more positive one than the existing militarization of borders by rich Western nations to repel climate change refugees. Achieving global sustainability is dependent on securing a long-lasting beneficial exchange between regions and people at risk and regions and people who live with abundance. Besides the maps that redraw the world into climate-affected and less-affected regions and countries without any reference to the territorial boundaries of countries, a new image of the world emerges revealing the porous geography of climate change. This new cartography of the world becomes not one of division but a running sequence of landscapes in degrees of environmental devastation aimed at provoking global direct action. The success of obtaining global sustainability is dependent on how the world can be envisioned to show a clearer picture of threatened environments and the people at risk. To make climate change 'globally owned' is to present a visual analysis of atmospheric weather turbulence and its effects on regions and people on the ground.

Mapping degraded and fragile regions to show adverse changes, such as soil degradation, water scarcity, health of plant and animal life, becomes a fluid rather than static process of representing the changing conditions on the ground. Global sustainability is then informed by intrinsic cartographical knowledge in tracking climate change to allow strategic policy responses to emerge. Shifting in scale from continents and countries to regions, towns, and villages, the cartographical maps can inform land care planning such as resting overgrazed/farmed areas to regenerate the soil and direct financial resources for land restoration and alternative forms of income. In many parts of the world, land care regeneration is being undertaken but in terms of scale it falls behind the current extent of land, forest, and sea devastation. Land and forest protectors such as the indigenous original owners and policing authorities are halting the advancement of deforestation by illegal loggers and herders. In some areas of the Amazon, CAR, and DRC in Africa, such initiatives are in place but again they are far smaller than what is required. Creating such intricate mapping information and directing policy management requires dedicated interdisciplinary and multinational research teams to gather on-the-ground analysis that can be processed, reported, and applied to a new international body dedicated to reversing land, forest, ocean, and sea ice degradation. Again, such bodies exist at the United Nations and various environmental foundations but far more executive powers and resources must be given to harness a truly global response to help regions most affected by climate change. Redrawing the world in areas of fragility and resistance, degradation and regeneration, productivity and financial resources without the constraints of sovereign nation-state boundaries would reveal how sustainability practices can be global by being regional.

Another map to be drawn would be a real-time depiction of land degradation from mining companies' resource extraction and where such resources are exported to fuel manufacturing, petroleum, and chemical industries. Showing and making accessible to people across the world through global telecommunication networks what has been hidden for so long in the gross environmental damage caused by mining and fossil fuel companies will provide information, and such information gives people the power to demand change. Itemizing the immense profits made from mining and fossil fuel companies and the costs of their operations in degrading whole regions and beyond, the earth's health can stoke a far wider militancy for appropriate taxation on these companies' immense profits. Instead of the large percentage of their profits going to shareholders and investors, it would be used to finance the restoration of degraded land and polluted lakes, rivers, and oceans caused by their operations and provide work for local people. The present global financial system will be subjected to the value of global sustainability and not the other way around where the present capitalist model benefits the richest 1% of the world's population who hold 45% of global financial value. The wealth of this tiny minority of the world's population was gained by utilizing the power of historical cartography that divided and dissected people, homelands, territories, and racial suppression instigated global inequality.

Substituting nation-state borders to regional-based mappings detailing the effects of climate change can bring a new way of perceiving the world in the 21st century. Where cartography illustrates the geography of a region through characteristic representations of the terrain, it can also bend the spatial perceptions of maps to show factual representations. In *Rethinking the Power of Maps*, Denis Wood refers to Gastner, Shalizi, and Newman's spatio-maps of state-by-state voting representation compared with population size of the 2004 US Bush/Gore presidential election as a spatial realignment mirroring reality. 'Where maps mirrors of reality this would have been an uncontestable conclusion, but maps are propositions—that is, they are statements that affirm or deny the existence of something—and alternative propositions were advanced immediately'.[5] Wood draws on Gastner and his colleague's political mappings to turn the spatial representations of cartography on its head and give a new physical actuality between the land mass of states in relation to voter population and in doing so configure a truer reality of the Bush/Gore presidential election.[6]

For the new cartographical maps to be influential as much as sensical, they must be filled with the intrinsic knowledge of regions and people. The maps are not aimed at radicalizing the cartographical maps of the colonial period, but instead will use that history to radicalize the effectiveness of climate change policies. Designed for a global audience where the information is readily accessible to the public, the cartograms show the cause-and-effect of climate change from those countries most responsible to those most vulnerable. Visually representing individual country sizes and populations in relation to their CO_2 emissions and the populations most affected from these emissions, the cartograms would swell and contract in response, showing the imbalance of climate change around the world. Where climatic turbulence in the atmosphere is oblivious to continents, countries, and regions, the climate change cartograms assert the global disproportionate values of CO_2 emissions between countries responsible and countries suffering its effects. More possibilities come to the surface in how cartographical maps can make sense of climate change and make it readily understood. For instance, a map showing once fertile lands lost to desertification and drought in relation to the dangerous routes that climate change migrants take to seek new opportunities to secure their lives illustrates the direct connection to the reality of climate change. Once again, the work of Decolonial Atlas has produced a substantial number of cartograms to build a picture of the world in terms of climate change, inequality, and land and sea degradation among others. Laying out their cartograms on a large table would show the earth in environmental turmoil, evidence of resource profiteering, land degradation, corruption and the disproportionate divisions between rich and poor countries, CO_2 emissions, and climate change. It might provide evidence for global justice to be finally served on those doing the most damage to the health of the earth and future generations.

To succinctly set out the value of cartography as proposed here, the aforementioned list provides the sequence of actions to build a new global picture.

- Cartogram illustrating territorial divisions of countries overlaid by adversely affected regions due to climate change that are geographically represented, not territorially.
- Cartogram illustrating degraded and fragile regions showing soil degradation, water scarcity, health of plant and animal life.
- Cartogram of climatic turbulence showing recurring extreme weather events year-by-year such as drought, flooding, hurricanes, heatwaves, ice melt.
- Cartogram combining degraded and fragile regions lost to desertification and drought in relation to the routes that climate change migrants take to seek opportunities for their survival.
- Cartogram of CO_2 emissions from countries worldwide to size of population and land mass.
- Cartogram displaying the global financial monetary system in relation to wealth generation and retention, e.g. the richest 1% of the world's population's share of global wealth compared with that of the majority of the poor.

What cartography can do is gather and visualize information to target programmes of direct resistance to the effects of climate change. Representing climate change damage to regions can mobilize specific actions such as water catchment, renewable energy production, drought-resistant crops, agricultural diversification, and soil restoration. Visualizing threatened ecologies through cartography allows for better policy formation and direct implementation so that adaptation can proceed rather than the abandonment of homelands and the displacement of tens of millions of people. For worldwide sustainability to be viable, it has to work at the regional level, restoring endangered ecologies and reclaiming agricultural land via global finances from fuels, taxing fossil fuel corporations' profits. A large part of this strategy is dependent on forming global responsibility to reform institutions and the private sector to give the world's 3.4 billion people affected by climate change alternative opportunities to support their lives. In their introduction to *Climate, Capitalism and Communities*, Astrid Stensrud and Thomas Hylland Eriksen argue that climate change strategies require reflexive action to have any real effect. 'Since human beings do not merely 'adapt to' changes in the environment, it must be assumed, at the outset, that people appropriate, engage, interact and negotiate reflexively with their environmental surroundings, as well as with markets, states, corporations and other powerful institutions'.[7] Drought, flooding, and water scarcity are not only a problem for regions and their people, but for the whole world—including those countries most responsible for the total percentage of global carbon emissions. As with water that follows the path of least resistance by abiding the formations of the natural terrain, likewise the future course of sustainability must be constructed by pathways of least resistance to counter climate change. The present failure of global sustainable policies in reaching CO_2 emissions reduction targets can be attributed to the political cycles of governments, global capital, multi-nationals, and the power of fossil fuel corporations' recklessness and avoidance of accountability. As the present fractured policies for global sustainability are being controlled by these forces, the focus should be on shifting the status quo where present wealth and power in the hands of a tiny minority cannot continue to dictate the majority, health, and cost of the planet.

Creating a global shift in how climate change can be represented such as productive regions (climate safe), non-productive regions (climate affected), remediated regions (land/ocean restoration), and climate refugees (migration routes) as a set of cartograms will give a fuller picture of what is unfolding across the world than the fragmented event-by-event recordings. It also offers the opportunity for this information to be 'transmitted electronically', as Nietschmann notes, to people throughout the world and across the territorial borders of separation. Specifically, it uses information to present a world picture of the power of capital, greed and corruption, corporate imperialism, industrialization, consumption, and the ongoing remnants of colonialism, which have shaped the world and are shaping climate change now. The dissemination of the cartograms would also place the health and protection of animal, plant life, and the environment at the core of sustainability and equality in people's hands—information is power. The present global conflict to reduce fossil fuel consumption to reach net zero carbon emissions is countered by the actions of global capital interests fuelled by powerful fossil fuel oil, trade, and energy corporations and governments unwilling to take decisive steps. The collective lack of ambition and short-sightedness to strategize long-term sustainable energy production cannot continue to be controlled by the capital power of the above actors. Humanity's protracted plundering of the world's resources only serves the profiteering of the minority, while the world's ecology and the majority of human, animal, and plant life becomes more endangered. With the integrated information the cartograms become powerful tools to counter governments and corporations through the dissemination of information. Openly illustrating their contribution to climate change and its effects on regions and their people can deliver decisive strategies for actioning and achieving global sustainability.

Climate Moves—Migrating Sustainability

Ground moves as we move over it, creating an intrinsic cartography of body and land and an imprint of the journey. The intrinsic cartography of body and land would unwrite itself in the human transition from nomadic life to permanent settlement to finally be expelled as mobility freedom gave way to claims to territorial lands and boundaries. Chapter 3 outlined this shift from the human experience of ground to shaping terrain for the extraction of resources and the sealed surfaces of cities and highways making it possible to move without touching ground. In *Dark Writing*, Paul Carter speaks of ground constantly moving, reforming our experience of movement: 'we do not walk with the surface', he writes, 'instead, we glide over it...in any direction without hindrance'.[8] Flattening, carving, and shaping terrain brought ground under control and mobile via threaded tarred ribbons of highways spread humanity's reach not through a connectedness with geography but a mental separation from it. In *Psychogeography*, Merlin Coverley locates this separation as 'the point at which psychology and geography collide, a means of exploring the behavioural impact of urban place'.[9] From settlement to mass urbanization, the collision between the natural world and human experience would create a disorder to these relations, the outcome of which has led to global environmental damage.

In the face of dissecting, flattening, and speeding across terrain without touching the ground, it was always going to be difficult for humanity to reimagine its way with the earth and build a global sustainable future. Dissolving this most fundamental relationship with the earth is implausible, it would seem, but the rapid expansion of urban centres has made visual ground invisible to the artificial surfaces of the built environment. The devastating atmospheric turbulence and on-ground environmental shock of global warming and climate change have shed a new light on humanity's relationship to nature and the necessity to remove the artificial surfaces dividing human experience from the natural world. One possible step is to break the ground of cities so the environs become porous to absorb the extremes of weather turbulence. Building-in rather than building-out weather would reconnect city dwellers to manage weather extremes. It should be understood that achieving sustainability must begin with increasing human exposure to the effects of human impact on the earth to radically change human behaviour. If such a course is not taken, where ground and weather connectivity are not integral to living in cities, then a large proportion of the world's population will remain disassociated from global climate change and from rural populations who are most directly experiencing its effects. To learn to live with the ongoing climate crisis is to break where 'psychology and geography collide' so as to form new flexible environmental adaptations as a part of living under extreme weather events in the 21st century.

As explored in Chapter 3, humans' superiority over their surroundings cultivated a new perspective towards the natural environment. Through

Figure 5.3 Highway overpass, Mexico City, 2021.

Source: Image by author.

the advent of agriculture this new perspective reduced the innate role of the natural world as inherent to human survival. Social theorist, Edward Soja, writes that humanity's ability to remove topography to build a new geography in its place is where '[a]n already-made geography sets the stage while the wilful making of history dictates the action and defines the story line'.[10] To rewrite humanity's recent history of expelling ground from its experience in constructing its cities is to create the deconstructive narrative of avoidance from the natural world, which half of the world's population live by. The story in rural regions is altogether different—theirs is one of non-avoidance. In many rural regions, people are being forced to leave their homelands devastated by the effects of climate change. Having to radically alter their lives, the once fertile ground they cultivated now expels them into migrant bodies in search of a new liveable place. Climate change in the 21st century has so far forced millions of people from their lands, at the beginning of a new epoch of mass human migration across the earth. From the human exodus out of Africa 100,000 years ago, to European migration to the Americas, Asia, and Oceania beginning in the 16th century, to enslaved migration during the colonial era of the 18th and 19th centuries, to the

Figure 5.4 Desert highway, aerial drone shot, NEOM region, 2022.

Source: Image courtesy of Otskydrone.

migration waves out of post-war Europe in the 20th century, humans have spread across the world's continents first through freedom of movement, then by enforcement, and now by climate change.

For sustainability to be a shared commitment connecting all people across the world, migration has to become an inherent part of living with climate change. Territorial boundaries denoting countries and government policies enforcing protectionism that divide the world and spatial definitions such as 'Global North' and 'Global South' can no longer dominate the order of the world if the collective goal of reaching global sustainability is to be achieved. In nomadic life, the wanderer moved with the ground they passed over, gathering knowledge of the terrain. The effects of climate change require a new set of knowledges—one for living with radically altered terrain and another to form the new passages of human migration where each step taken unearths new possibilities for environmental adaptation. The present response has seen millions of people having to leave their devastated homelands and move to cities unable to cope with the mass influx of new rural migrants and further to entirely new countries far from their homelands. As previously cited, the effect on people living in rural regions devastated by climate change while increasing exponentially is unsustainable given that by 2050, 70% of the global population, which is predicted to be 10 billion people, will live in cities. The immense environmental devastation from climate change and the scale of human migration are clear signs of the collision course on which humans and the earth are on. The mass abandonment of rural areas due to the effects of climate change, lack of opportunities and jobs is by all accounts unsustainable. Sustainability is not just about finding ways to counteract climate change, it is also about counteracting mass human migration into cities. Abandoning rural regions and the subsequent reduction in the number of people cultivating the land is no doubt alarming to the future of humanity's capacity to feed itself, restore degraded lands, and create adaptable living in response to climate change. The problem increases when acknowledging that cities have centralized capital and global governance, industrialized the production of goods, services, and labour, and importantly maintained the huge wealth gap between urban and rural populations. How human life on the earth can be reformed is to ensure that sustainability addresses the power of cities, the adaptability of its inhabitants, and the mobility of rural populations to expand human diversity in living with climate change.

Modern migration is restricted by the territorial borders of countries but if human migration is an inescapable part of living under climate change and achieving global sustainability, then the exclusive territorial zones of nations must be renegotiated. One way is to begin the shift away from centuries of human migration into cities back to rural areas, such that abilities for adaptation to climate change can increase and diversify. At present it is the rural populations of farmers and villages who are most vulnerable to the effects of global warming, but who are also possibly more adaptable and capable of defending themselves from climate change than the defenceless urban dwellers who rely entirely on the importation of food and resources. The static entity of the city means it is less able to adapt and be self-sufficient in the face of the challenges of climate change. Programming the cross-flow of migrants between regions and between cities and regions increases self-reliance, for it places people at the centre of changing human adaptation to climate change rather than dreaming of technological solutions to solve human impact on the earth.

Human mobility and connectivity with ground as a liveable solution directly write sustainability into the human experience. Blurring the boundaries between urban and rural, territories and borders, allows for a new evolution of human migration to climate change as a positive step rather than the present enforced step. Shifting finances away from the city to rural areas increases the degree and diversity of human adaptation to changing environments. Challenging international boundaries that propagate territorial division and shifting individual mindsets concerning national identity will require real change to create a global collectivism that connects sustainability directly to ground, mobility, nature, and the human experience. Up until now it has been possible for city dwellers to deflect the effects of climate change, but this cannot continue given the increases and regularity of hurricanes, cyclones, extreme temperatures, rising sea levels, and the energy requirements to run cities.

Humankind invented technologies to dominate the earth's natural environments but now find themselves fearing these inventions. With excessive resource extraction, manufacturing and consumption of goods, relentless growth and CO_2 pollution of the atmosphere, humanity is now more exposed to the elements of the earth than ever before. City inhabitants have become captives of their constructed environment, and rural populations are captive to the effects of extreme weather events. To resolve these divisions and experiences between city dwellers and land cultivators is to link them to a new cultivation of the earth. Where agriculture revolutionized humanity's relation with ground with the invention of the plough, a new evolution of climate change migration will revolutionize human habitation on the earth. To decrease land degradation and increase land cultivators is at heart of this revolution. The plough that furrowed the soil and secured food production for human survival can be reapplied in the next steps to cultivate new relations between humans and the natural environment as climate change progresses in the 21st century. To reconnect humans with ground is not only to physically re-embed ground into the human experience, it is also to unearth a new evolutionary psychogeography of the mind. Within this physical shift of human embodiment to ground moves and migration, the impetus for sustainability is to inform new modes of living. Ways of contesting the human impact on the earth, through actions such as reducing consumption, abandoning fossil fuels, and recycling, can ultimately only be successful if accompanied by a whole-Earth living practice within the body, ground, and the natural world. The present gap separating humans and ground, especially those living in cities, is only held together by the thinness of the artificial surfaces and psychological mindset that divides them. Unearthing the ground that divides humans, walking terrain, and reinstalling weather into their daily experience opens the space for humans to comprehend nature, animals, plants, and all living things as equal entities in their pursuit of survival.

Human migration cannot go on being viewed as a threat to the territorial boundaries of nation-states. Instead, human migration must be supported as an effective management in responding to and living with climate change in the 21st century. Free movement is a human right, having to abandon homelands unable to support life cannot be criminalized under the global condition of climate change. Retracting the freedom of movement by forcing people to remain in regions unable to support them illustrates the continuing collapse of human-to-human ethical responsibility in the face of global warming. The present engrained economic inequality between countries and continents initiated under European invasion, colonialism, stolen resources, and industrialization is now also repeated between those who bear the brunt of the catastrophic effects of climate change (many of which are former colonial countries) and those who contribute most to the world's CO_2 emissions (rich industrialized nations—a portion of former colonialist coun-tries) but who experience far less the damaging effects on their livelihoods or standard of living. In a divided world of militarized territorial boundaries and the rich nations' fear of being overrun by climate change refugees of different racial backgrounds, religions, and cultures can only set the world on a course of increased global conflict.

There is an opportunity to avert such conflicts by incorporating climate change migrants into a new global monetary system that crosses existing cartographical divisions to create inter-regional, inter-national, and inter-continental economic provinces. Ownership of productive fertile land regions capable of supporting small intense cultivation by farmers who have had to abandon their lands would no doubt need to be financed. The cost of pre-paring and accommodating these farmers would far outweigh the problems and the conflicts that will continue to grow if they are not assisted. Such a model challenges the foundations of nation-states, which are based on seg-regation and the exclusion of 'unwanted' migration. But as climate change increases, such protectionism becomes an ever-more archaic and inflex-ible model. Accommodating new populations of people to settle in regions spanning countries and acquiring the land in which they can cultivate and provide food security for themselves is creating sustainable lives while con-tributing to a region's economy. The challenge facing rich Western coun-tries who have gained most from extracting the earth's resources is either the incorporation of a new land distribution and economic model for climate refugees or a continuation of an unsustainable enforcement to repel them from their borders. Deploying nationalism and patriotism as tools to manu-facture fear and threat as the means for sustaining living standards is what drives rich Western nations' greed to not accommodate people who have lost everything due to climate change, a disaster that was not of their making.

The historical scars, marks, and lines of the cartographic maps divided continents and imported ownership by imperial nations cannot continue to dominate under the global climate crisis the world is now experiencing. Suppression remains a key motive by rich Western nations to prevent the migration of climate refugees, and drawn borders separating countries and peoples continue to support oppressive regimes of global division. Keeping intact the protracted separations between the 'Global North' and the 'Global South' perpetuates the present system of inequality and differential experiences of the effects of climate change and, importantly, it dissipates the ability to realize global sustainable policies. The environmental destruc-tion caused by drought, land degradation, desertification, and flooding that threatens habitats, ecologies, and human survival is not a selective one but a global movement that shifts between regions, countries, and continents. The unpredictability of weather turbulence due to climate change calls for a new human experience to the natural world—the removal of surfaces and divisions that separate rural and urban populations and a redrawing of the global cartographical map to climate-affected regions. As global sustainable policies and reductions in greenhouse gas emissions targets to combat global warming fluctuate, human-to-human and human-to-Earth disruption will continue to grow, resulting in a systemic failure of sustainability in the face of the global migration of people seeking new lands and opportunities for their survival.

Mobile Ecologies—Environmental Costuming

In his essay 'The Viable Human' from the book *Deep Ecology for the Twenty-first Century*, ecotheologian Thomas Berry advocates a complete rethinking of humanity's habitation on the earth based on establishing a set of values for its future existence. He argues that averting the catastrophic collapse of the world's ecosystem is contingent on valuing how 'each individual being is also supported by every other being in the earth community. In turn each contributes to the well-being of every other'.[11] Berry's advocacy seems simple enough to grasp though cynics would have a field day given humanity's propensity for conflict. Asking humans to unite under a common cause calls for a radical realignment in human trust and compassion. In the preface to the same collection, George Sessions describes deep ecology as a paradigm shift in human history: 'to protect the planet from ecological destruction involves the move from an anthropocentric to a spiritual/ecocentric value orientation'.[12] The term 'deep ecology' was first coined by Norwegian philosopher, Arne Naess, in his 1973 article 'The Shallow and the Deep, Long Range Ecology Movements: A Summary'.[13] Naess's primary approach to deep ecology was to instill a philosophical societal shift 'where the question is "deep" and public' in order to move nature on to an equal footing with humanity.

Realigning humanity's ecocentric place in the world with all living things certainly seems like a utopian ideology given the history of humanity. Creating words and terms and trusting in them to convey real meaning in tackling climate change is less guaranteed when civic society is idealized. During the 27 COP meetings on climate change since 1995, an endless array of terminologies and project titles for actioning global climate change strategies have suffered when colliding with the actuality of their attempted implementation. Many climate change solutions humans have implemented to synchronize their 'deep ecology' has entailed focusing on technological solutions rather than addressing wealth distribution, excessive extraction of resources, and consumption. Geoengineering initiatives—from capturing greenhouse gas emissions from industry, carbon-capturing plants, genetically modifying farm animals to reduce methane gasses to proposing atmospheric discs to deflect solar rays from the sun—are trundled out as solutions to the climate crisis. These solutions essentially allow for humanity, especially rich Western countries, to continue their present way of life in the belief that technological innovation will prevail in averting total environmental collapse.

While there has been a global shift in the awareness to climate change, a general attitude still persists to continue living as normal and to accept a universal human condition of global inequality, poverty, racial discrimination, ideological, cultural, religious, and gender suppression. Berry's philosophy where each individual 'contributes to the well-being of every other' seems far removed from the realms of the possible when the ideology for a 'world-wide collective well-being' is daily eroded on all sides of the political spectrum from autocracies to theocracies and even democracies. Berry talks of the rights of the natural world 'still at the mercy of the modern industrial corporation as the ultimate expression of patriarchal dominance over the entire planetary process. The four basic patriarchal oppressions are rulers over people, men over women, possessors over nonpossessors, and humans over nature'.[14] Rethinking human conflict and dominance over nature is part and parcel of dismantling the systems of division and for sure this is not going to happen any time soon. 'Through modern technology we can manufacture great quantities of products with greater facility', Berry writes. 'Human technology also enables us to travel faster and with greater ease. So, on and on, endlessly, we see our increasing human advantage over the natural world' and, it should be added, between ourselves.[15]

In another essay from the same book, entitled 'Deep Ecology: A New Paradigm', Fritjof Capra writes that forming a deep ecology with the earth 'does not separate humans from the natural environment, nor does it separate anything else from it. It does not see the world as a collection of isolated objects but rather as a network of phenomena that are fundamentally connected and interdependent'. Deep ecology, Capra assures us, 'recognizes the intrinsic values of all living beings and views humans as just one particular strand in the web of life'.[16] He also argues that it is 'most important to introduce ecologically orientated ethical standards into modern science and technology'. While well-meaning, the parameters of ethical standards are constantly shifting especially in light of the rapid rise in use of artificial intelligence across all areas of human society. Applying an ethical model or moral code seems somewhat irrelevant when the ethics and morality are mostly absent from governments' and corporations' drive to accelerate economic growth, resource extraction, and profit while inadvertently maintaining racial and global inequality in the process. Strategizing sustainability is not about deploying 'what is right', it has to take on a far more militant role that can rival the environmental aggression and power of global capital system, governments, and corporations. Applying ethical and moral sensibilities to seek to shift humanity from its historical roots of conflict and dominant position over all other living things on the earth, whilst laudable in its motives, stretches belief in humanity's capacity to 'work together' beyond credibility. The fluidity of ethical and moral codes is written as needed and applied when suited; in other words, placing codes of trust in human behaviour given the history of proven failure is naïve and dangerous. So entrenched are the forces of authority and discrimination on the earth, humans have come live with acts of cruelty and destruction. To place ethical and moral behaviour as a way for humanity to tackle climate change and the issues surrounding it would be hard to justify.

Deep ecology is not simply about connecting human external relations to the natural world but also the internal relations within it. This is not to be confused with the forging of a spiritual dimension; rather it is a functional embodiment of ecology at the centre of human existence in the 21st century. Returning to John Berger's 'Why Look at Animals?', explored in Chapter 4, the answer is for humans to comprehend their surroundings in connection to how animals comprehend theirs. Animals, Berger asserts, are inseparable from their surroundings and as such there is no distinction between them and their environment. Humans have progressed to comprehend themselves as infinitely separable from animals and the wider natural world and thus they are able to form their relational indifference towards both. While this is not applicable to first nations and indigenous peoples who still live by their deep ecological connections to their surroundings, it certainly applies to a great part of the human population. It is fundamental to understand how human indifference to the natural world has led to climate change—it simply made the progressions to industrial-scale environmental destruction easy. Where animal and plant life succeeded in creating their diversity of appearances and behaviours through environmental adaptation and behaviour, humans reduced theirs in order to form a distinctive distance. The impenetrable barrier humanity has formed between itself and everything else on the earth has come at the cost of reflecting it into its own behaviour. So fast has human progress been since the beginning of industrialization 250 years ago that a crisis of identity within the world has emerged. This crisis is no longer the confrontation with the natural world, the crisis is now what to do if humanity were to give up its control over the earth. Humans are having to rethink who they are in relation to the earth and what they have created, which is to say that who humans are is measurable by what they have destroyed.

Humans set their course on the earth to be entirely different from the living and non-living. This is what has defined humanity and secured its survival. The crisis can be traced back to the beginning of settlement to the mass production of objects and speed of transportation, as already stated, that cast the earth to the periphery and centred humanity in its place. The crisis can also be the product of humanity's ability to conceive ideas, design, and produce from what the earth supplies rather than master an inclusive relation to the natural world. In its drive for progress, humanity has been unable to define another way of how to live on the planet other than through conflict and war, extraction and consumption of resources, greed and corruption. Humanity needs to think what it wants to be and how it can achieve it. The catastrophic effects of climate change have made this new search all the more pressing, for humanity is aware that the decisions it makes now will have profound ramifications for the future generations and well-being of the earth. The climate crisis is challenging human behaviour, which until recently went along unchallenged and now humanity is afraid (of itself), faced with the growing knowledge of what it has been doing to the planet. Acknowledging the climate crisis is one thing; however, keeping growth and capital in the same centralized position to human life on the earth is to say: 'let the future generations work it out'. Humanity is in a place of great insecurity. The climate crisis in the 21st century is indisputable and will either define humanity's capacity and intelligence to address it or confirm its stupidity and inability to cooperate. Where governments and corporations still persist in maintaining the global status quo and place profit before the health of the earth, humans' downfall seems imminent.

Change in individual human behaviour can overcome governments and corporations to bring about wider change in societal and cultural behaviour. For humanity to get to a place where it is effectually creating change in how it relates to the earth, it has to dismantle some parts and adapt others to create something new. It is not the case of reverse engineering all the problems humans have created on the world's ecologies, but to reverse what has driven them up until now in terms of wealth, growth, and success. How far humanity must 'go' to change itself in order to change the world is an 'unmeasurable known', but taking into account what it has been able to do in the last 250 years should be a marker of what *not* to do, and not only in terms of the next 250 years but the next 50 years, such is the compressed timeline of climate change action humanity is facing. Climate action groups such as Extinction Rebellion, Just Stop Oil, and Last Generation, who carry out militant actions to immediately bring a halt to the use of fossil fuels, are models of global action—theirs is not an ethical or moral stance but one of absolute necessity. Yet instead of being encouraged and congratulated, they are derided and criminalized. Taking their measure to a global scale would have an immediate effect on business, transport, manufacturing, and energy production with most, if not all, coming to a standstill. It would certainly force humans to adjust to a very different world where the forces of power and capital have diminished to the betterment of the earth. It would force humans to adapt to a radical shift to consume fewer products and goods and to make do with less comfort. It would make people less wasteful, more resourceful, and more reliant on each other. Giving up a 'lot of stuff' that humans consume but do not necessarily need, sharing resources and appliances, and finding better ways to merge with the ecology goes some way to confront the emergency humanity faces but at the same time also stalls the rebellious action required to radically change human habitation on the earth.

At present global sustainability policies do not pursue such a radical agenda that is essentially based on de-capitalizing the world financial system, bringing down global inequality, and repairing the histories of colonial suppression and present-day corporate imperialism. To be sure, what Just Stop Oil, Last Generation, and many other climate change action organizations want is not going to happen in the short or immediate term or, disastrously, if at all. For sure, governments and business do not want to see societies operating independently from them and will make assurances that such radical steps that would mean dissolving the powerful and the rich for the common good of all people across the world will not happen and that any attempt to do so will be brutally put down. Yet, the risks we are talking about affect the top 10% of the world's population who own 80% of global wealth. For the other 90% of the world's population, it would mean relieving poverty and redistributing wealth and with that a shared cooperation concerning the future direction of human life on the earth. At present the position taken by sections of government and industry is to advocate continued growth and to put their trust in the uptake of renewable energy as the key to averting further climate devastation. While some groups advocate technological innovation to power humanity out of the crisis, others argue for depopulation and management of human life. No doubt the climate crisis threatening the world is a crisis in the capacity of human consciousness to comprehend its own destruction. Scientists, governments, businesses, and societies know what to do but there is a whole swathe of obstacles that delay real decisions. This is not lost for example in the new nuclear arms race presently unfolding and the rise of human conflicts across the world that continue to jettison any advances in achieving global sustainability. The proliferation of tactical nuclear weapons coupled with dangerous rhetoric and propaganda from Russian, North Korean, and Iranian governments and Western nuclear countries to counteract would seem to point to the inevitability of nuclear war and, while alarming, it is characteristically, destructively, all too human.

Over millennia, societal change against the ruling classes has come as a result of struggles (both violent and peaceful) to remove oppressive rulers, tackle injustices, enforce rights, remove discrimination, and more. Uprisings and revolutions modified societies and forged civil societies and no doubt it will be the same as the climate crisis deepens by a rising global populace seeking to challenge the power of capital and reverse environmental destruction. There is increasing momentum behind the call to abandon the goal of wealth creation that has driven humans' desire for a new ecological global society to be born. Economic divisions and inequality and histories of persecution and greed between people, races, and cultures make the success of a radical change a difficult enterprise to undertake, yet it is the only enterprise humanity has to change its course. To wait for change to happen is to remain beholden to the wheels of capital and corporate power whose interests are in maintaining growth and profit for the minority. Creating ecologically sustainable societies means disassembling the present forces that control human life on the earth. All human life and activity releases disruptive atmospheric and ecological damage—this is unavoidable—but as Chapter 4 set out, the future of human environmental security entails a closer connection to animal and plant life to form an ecology of coexistence.

Reversing resource exploitation and environmental degradation rests on individuals to conceive their ecology and adaptation for 'wearing our ecology'. Writing back in 1985, Bill Devall and George Sessions, in an earlier book simply titled *Deep Ecology*, advocated an ecotopian vision as 'affirmations of our bonds with Earth'. 'Creating ecotopian futures has practical value. It helps us articulate our goals and presents an ideal which may never be completely realized but which keeps us focused on the ideal'. They suggest that 'ecotopian visions give perspective on the vain-glorious illusions of both revolutionary leaders and the propaganda of defenders of the status quo'.[17] To avoid illusionary ideologies in the face of powerful 'defenders of the status quo', the term 'wearing your ecology' is used to apply a visual element for human adaptation to climate change. While sustainability is the end goal to halt environmental damage, dismantling capital and greed, it requires preparation to install universal cooperative living with the earth's ecologies. To 'wear your ecology' is to comprehend the natural environment as intrinsically human. To governments and corporations, it would spell their demise in controlling life on the earth and to environmentalists their advocacy for a passive human and Earth coexistence.

Present initiatives for global sustainability are based on the capital model where the world's wealthy nations create financial incentives for changes in consumption and energy production in their own countries and with extra money left over to fund adaptation schemes in the most vulnerable countries exposed to climate change. The 'Deep Ecology' proposed by Devall and Sessions is based on a self-realization where 'the insight draws us to respect all human and non-human individuals in their own right as parts of the whole without feeling the need to set up hierarchies of species with humans at the top'. This view has a bearing on realigning humanity's understanding of itself.[18] Likewise, 'wearing our ecology' is to come to a point of 'biocentric equality' where 'there are no boundaries and everything is interrelated'. Forwarding environmental human costuming is transferable if we think of how humanity has artificially 'costumed' its way of life in the building of cities and devising technologies that have separated human experience from the natural environment—it is a matter of relearning. 'Wearing our ecology' is reachable for the transition required for humans to *become* their surroundings is to regain their (natural) instincts—the instincts humans reengineered to *become* inseparable from the city. 'Wearing our ecology' is experiential in the making, evolving, and shaping of human life to nature. It is classless, equitable, without racial definition, religious and gender inclusive, that is, the opposite of the economic and cultural divisions and the way humans think now. How to be effective and impress upon meaning in shifting human activity on the earth is the work of all humans. As such 'wearing our ecology' is a shift in human activity on the earth for every human to create new ecologies with the earth and between each other, each one expanding on the previous. It is an extension of human progression that has gone before though it is radically different in its application.

Early humans lived by first coexisting with the natural world through threat and fear for their survival. Over hundreds of thousands of years, humans progressively built up their security through the crafting of tools, objects, and weapons that enabled them to rise to the top of the natural world. What is now evident is that this ascendency of humanity has been left unchecked and the massive expansion, consumption, and disregard of the planet has blown the earth's climatic balance off course. Humans have devised technologies of great complexity and at the same time they have devalued nature's complexity to assert control. Now the roles have been reversed as humans need to gain support of all other living things if their existence is to continue. They also know that they cannot keep on destroying natural environments for their own benefit and polluting the atmosphere, rivers, and seas as a result of processing the earth's resources. The choice of either continuing rampant consumption and environmental destruction or abandoning their desires is the conundrum now facing humanity amid climate change. 'Wearing our ecology' means to subsume nature within the body and daily life, and if that can be done then the divisions between nature and people and between rural and urban populations' experience of climate change will diminish. The forced nomadism of people from rural areas to cities and from the 'Global South' to the 'Global North' will be given pre-cedence as new frontier explorers in living with climate change, rather than being met with subjection and rejection. The dominance of the world's pre-sent polar division must be converted to where the eight billion people living on the earth have the equal right to live their ecology on and through their body and to allow nature to re-emerge as their equal.

This chapter has attempted to connect human and Earth coexistence to the notion of 'wearing our ecology' based on co-dependence with the natural world. In the era of climate emergency, the old cartographic lines of spa-tial delineations must be dissected, boundaries dissolved, artificial surfaces of the city made porous, and connections between rural and urban populations bridged for sustainability to become a global reality. A new cartography of climate change will begin to emerge across the continents foregrounding a new geo-spatial global order. At first the mass of lines criss-crossing the globe illustrating the mobility of climate refugees will stabilize to forge a new nomadic resettlement of the earth where individuals and collectives operate independently of established authorities to devise their own ecological soci-eties based on living with the least resistance on the earth.

Rethinking the present structures guiding our living conditions to con-ceive a reduced human impact through a wearable mobile ecology remains nascent in the global context. 'Wearing your ecology' requires a far greater connectedness between humans and the environment. It also means far greater human mobility across the world than exists now, opening new opportunities for transformation between urban and rural regions. It also means the removal of global divisions such as the 'Global South' and 'Global North'—terminology that only reinforces inequality. Adaptability is key but in order to achieve it, humanity must learn to become its ecology and comprehend itself as its surroundings. The next chapter 'Environmental

Adaptations—Spiral-Swarming, Human Diversity' explores how ideas and abstract conceptions for future human coexistence with the earth can be sited and augmented.

Notes

1 James, Corner. 'The Agency of Mapping: Speculation, Critique and Invention'. In *Mappings*, edited by Denis Cosgrove. London: Reaktion, 1999, p. 215.
2 See Edward W. Soja. 'Writing the City Spatially'. In *City: Analysis of Urban Trends, Culture, Theory, Policy Action*. London: Routledge, 2003, p. 275. In another publication Soja suggests that cartography is a type of illusionary device of 'an already-made geography' that 'sets the stage, while the wilful making of history dictates the action and defines the story line', which plays well into the European colonialist perspective, appropriation, and dissection of the African continent. See Edward W. Soja. *Postmodern Geographies: The Reassertion of Space in Critical Social Theory*. London: Verso, 1989, p. 14.
3 Adding to Nietschmann's quote collating guns with maps, he writes:

> whereas maps like guns must be accurate, they have the additional advantages that they are inexpensive, don't require a permit, can be openly carried and used, internationally neutralize the invader's one-sided legalistic claims, and can be duplicated and transmitted electronically which defies all borders, all pretexts, and all occupations.

See also Joe Bryan and Denis Wood. 'Maps, Guns, and Indigenous Peoples'. In *Weaponizing Maps: Indigenous Peoples and Counterinsurgency in the Americas*. New York: Guilford Press, 2015, pp. 74–95.
For a detailed analysis of conflict between nation, territorial boundary, and original indigenous owners, see Bernard Nietschmann. 'The Miskito Nation and the Geopolitics of Self-Determination'. *Journal of Political Science* 19, no. 1 (1991): Article 3: https://digitalcommons.coastal.edu/jops/vol19/iss1/3
4 To view the comprehensive mapping undertaken by contributors to Decolonial Atlas, see: https://decolonialatlas.wordpress.com/
5 See Denis Wood. *Rethinking the Power of Maps*. New York: Guilford Press, 2010, p. 41.
6 The cartograms were created by Michael Gastner, Cosma Shalizi, and Mark Newman for the 2004 US Bush/Gore presidential election. The cartographers developed a spatial rendering of Democratic (blue) and Republican (red) votes cast in relationship to the population sizes of states, the electoral college system, and individual counties. The cartograms show an overwhelming US map dominated by the colour red (Republican) compared with a far smaller amount of blue (Democratic). But when the votes from each state are converted to individual states' populations, a very different saturation of red versus blue emerges. States expand and shrink in size to reveal that the dominance of red covering has reduced and the blue has increased, revealing a map that is far more balanced between the two colours. This is highlighted by looking at America's most populous and Democratic state of California, which has doubled its surface coverage of the map, while lesser populated states such as Wyoming have shrunk in proportion to its

small population compared with its size. The cartograms not only swelled and contracted the natural boundaries of the states to illustrate a very different spatial representation of American voters and state populations but also reflected the spatial concentrations of the liberal west and east coast and north central states in comparison with the conservative middle and southern states. See www-personal. umich.edu/~mejn/election/2004/

7 Astrid Stensrud and Thomas Hylland Eriksen, eds. *Climate, Capitalism and Communities*. London: Pluto Press, 2019, p. 3.

8 Paul Carter. *Dark Writing: Geography, Performance, Design*. Hawai'i: University of Hawai'i Press, 2009, p. 6.

9 Merlin Coverley. *Psychogeography*. Hertfordshire: Pocket Essentials, 2006, p. 10.

10 Soja, *Postmodern Geographies*, p. 14.

11 See Thomas Berry. 'The Viable Human'. In *Deep Ecology for the Twenty-first Century*, edited by George Sessions. Boston: Shambhala Publications, 1995, p. 13.

12 Ibid., p. 21.

13 See Arne Naess. 'The Shallow and the Deep, Long-range Ecology Movement. A Summary'. *Inquiry* 16, no. 1–4 (1973): 95–100.

14 Berry suggests that the patriarchal system has and is still dominating human life on earth:

> Gender has wide implications for our conception of the universe, the earth, and the life process, as well as for the relation of human individual toward each other and for identifying social roles. The industrial establishment is the extreme expression of a non-viable patriarchal tradition.
>
> (Berry. 'The Viable Human', p. 14)

15 Ibid., p. 13.

16 See Fritjof Capra. 'Deep Ecology: A New Paradigm'. In *Deep Ecology for the Twenty-first Century*, p. 20.

17 See Bill Devall and George Sessions. *Deep Ecology*. Utah: Peregrine Smith Books, 1985, p. 162.

18 Devall and Sessions suggest that humans have existed on far more material needs than necessary.

> Our material needs are probably more simple than many realize. In technocratic-industrial societies there is an overwhelming propaganda and advertising which encourages false needs and destructive desires designed to foster increased production and consumption of goods. Most of this actually diverts us from facing reality in an objective way from beginning the 'real work' of spiritual growth and maturity.
>
> (Ibid., p. 68)

Bibliography

Berry, Thomas. 'The Viable Human'. In *Deep Ecology for the Twenty-First Century*, edited by George Sessions. Boston, MA: Shambhala Publications, 1995, pp. 8–18.

Capra, Fritjof. 'Deep Ecology: A New Paradigm'. In *Deep Ecology for the Twenty-First Century*, edited by George Sessions. Boston, MA: Shambhala Publications, 1995, pp. 19–25.

Carter, Paul. *Dark Writing: Geography, Performance, Design*. Hawai'i: University of Hawai'i Press, 2009.

Corner, James. 'The Agency of Mapping: Speculation, Critique and Invention'. In *Mappings*, edited by Denis Cosgrove. London: Reaktion, 1999, pp. 213–252.

Coverley, Merlin. *Psychogeography*. Hertfordshire: Pocket Essentials, 2006.

Devall, Bill, and George, Sessions. *Deep Ecology*. Utah: Peregrine Smith Books, 1985.

Naess, Arne. 'The Shallow and the Deep, Long-range Ecology Movement. A Summary'. *Inquiry* 16, no. 1–4 (1973): 95–100.

Nietschmann, Bernard. 'The Miskito Nation and the Geopolitics of Self-Determination'. *Journal of Political Science* 19, no. 1 (1991): Article 3. https://digitalcommons.coastal.edu/jops/vol19/iss1/3

Soja, Edward, W. *Postmodern Geographies: The Reassertion of Space in Critical Social Theory*. London: Verso, 1989.

Soja, Edward, W. Writing the City Spatially'. 269–280. London: Routledge, 2003, pp. 269–280.

Stensrud, Astrid, and Thomas Hylland, Eriksen, eds. *Climate, Capitalism and Communities*. London: Pluto Press, 2019.

Wood, Denis. *Rethinking the Power of Maps*. New York: Guilford Press, 2010.

6 Environmental Adaptations
Spiral-Swarming, Human Diversity

Oscillations—The Climate Spiral

The previous chapter sought to redefine how colonialism used cartography to claim ownership over invaded lands and established flaws in the territorial division of land in contrast to the cultures of the people. Cartography subjected the characteristics of geography to two-dimensional representations in lines and marks, setting ground as static rather than mobile and constantly shifting as we pass over it. It presented the idea that human mobility will play an important role in environmental adaptation to the effects of climate change in the 21st century as early human migration across the world had done prior to the establishment of settlement. Finally, the chapter proposed the inception of a human–ecological connectivity for 'wearing our ecology' through gathering terrain, forging connections to the environment, animal and plant life, abandoning excessive consumption, and forging new alternative societies in sync with the natural world as one pathway for the future of human and Earth coexistence. This chapter, 'Environmental Adaptations', proposes alternative modes of living under climate change by asking how humanity might rethink its inhabitation on the earth.

The excesses of the 20th century's mass production, consumption, and ecological devastation now threaten humanity in the present century—one that is being defined as the era of catastrophic climate events. The alarm of this imminent threat was sounded in the late 1960s and 1970s by environmental scientists' findings regarding human impact on the earth's biosphere. Two decades later, their unheeded warnings were highlighted by American scientist, Carl Sagan's 1985 testimonial before the US Congressional Committee on the Greenhouse Effect. These early warnings were challenged, largely dismissed, and characterized as a thorn in the side of economic growth and as a consequence they were played down to the detriment of the earth's ecological health. Nearly two centuries before Sagan's testimony, the German botanist and geographer, Alexander von Humboldt, conceived his theory of comparative nature—a biological–botanic synthesis between all life on the earth in a unified relational system. As with the scientific warnings, Humboldt's relational Earth system had little effect when the ravages of mass

DOI: 10.4324/9781003382515-7

resource extraction and industrialization began to take hold in the second half of the 18th century. The environmental costs of negating the early signs of catastrophic environmental damage are now clear in the 21st century. Around the world, natural ecological systems are declining at a rate not seen in human history, and hundreds of millions of people are having to leave their homelands unable to support their lives. Humanity's plundering of the earth's resources over the last two centuries and dismissal of irrefutable scientific evidence and advice to reverse or at least curtail ongoing environmental devastation has come at its own peril.

Answers to counter the effects of climate change have so far been centred on eliminating fossil fuel dependency to reduce air pollution and in doing so reduce atmospheric turbulence that is driving catastrophic weather events. Immense 'clean-up' programmes have been introduced to decrease landfill waste, rid plastic pollution from the oceans, better manage natural resources, and battle against deforestation, desertification, soil acidification, and dwindling fresh water supplies. It was not until the 21st century that a real international campaign to address these issues became a matter of global importance. Large-scale sustainability programmes have been launched to intensify green energy production, electrification of transport, banning of single-use plastic, waste recycling management, and addressing global inequality. Much is yet to make an impact in face of the continual drive for economic growth and consumption. Many people are 'doing their bit' to regain Earth's climatic balance for the next generations, yet the often-voiced phrase 'time is running out' is being heard but somehow not universally believed.

Global CO_2 emissions are forecasted to rise over the next 20 years before falling as the impact of sustainability strategies come into effect but in that time a great deal of irreversible damage to the environment will have been done. The phasing out of non-renewable fossil fuel use and transitioning to carbon neutrality through wind, solar, and hydrogen energy generation and creating incentives for 'green living' are aimed to avert further increases in the planet's temperature. Other initiatives include developing drought-resistant crops through genetic seed modification, better water catchment in dry regions, land remediation, and global fish stock regulation. Yet, pressure on the earth's resources is increasing and as the present 8 billion people who inhabit the earth is projected to rise 10 billon by 2050, the actions in place now to decrease human environmental damage will surely be tested as a matter of course through population growth and expansion. Global sustainable policies are nowhere near the scale they need to be if they are to support sustainable food production, eliminate economic inequality and poverty, create equal opportunities, as well as contend with cultural issues such as racial, religious, and gender discrimination—all of which impact on the success of tackling climate change. If this were not enough, human conflicts such as the Russian war in Ukraine and many other worldwide conflicts place the ability to reach global sustainability in a fragile, if not impossible, position to be effective. The global list of severe environmental issues and catastrophic weather turbulence facing humanity has reached a precipice—a balancing act of being checked and spinning out of control.

To climb down from this metaphorical precipice is not helped by global bilateral relations becoming increasingly more polarized and unstable between democracies, autocracies, theocracies, and dictatorships making use of nationalism and patriotism for political point-scoring. The global imbalances of wealth and poverty, privilege and hardship are increasing rather than decreasing, and the human right of universal freedom for people to access the world is becoming ever-more constricted through the militarization of nation-state borders. Global sustainable policies have further divided the world, for they highlight the vast chasm between the vulnerable and protected populations. Much is being made of inter-governmental agreements concerning the global transition to green energy and sustainability. The agreements and resolutions, such as those adopted from the 27 COP conferences beginning in 1995 on climate change, have tended to highlight global commitments but the real work of financing them has not reached the amounts agreed upon. Governments swayed by populism and corporations driven by profits are trading economic growth and consumption for further environmental damage. Mediatized messages telling global citizens what's needed to be done to combat climate change are loud and clear as much as alternative messages are shrouded in foggy information by those willing to manipulate communication for their gain. Capitalism, corporate power, and political electoralism mean that the health of the planet and the lives of future generations are being gambled away. As the major contributors of global CO_2 emissions, wealthy nations are balancing ongoing emissions with maintaining their wealth and living standards at the expense of the poor, since it is the resources of poor countries that are sustaining the wealth of the rich countries. Countries such as the United States, China, India, Europe, the United Kingdom, Russia, Saudi Arabia, and Australia, to name a few of the major suppliers and contributors to CO_2 emissions, contrasts to many of the world's poorest countries who pay a far higher climate change cost in terms of their livelihoods. As previously cited, in proportional terms, countries on the African continent, possibly the most affected by climate change, account for only 3.7% of total global CO_2 emissions.[1] It is difficult to comprehend how a global pact between countries can emerge to create worldwide sustainability given this acute imbalance.

At the 2022 COP27 conference, wealthy nations made commitments to contribute to a funding package of $US100 billion dollars over 10 years to the Adaptation Fund to help finance the most vulnerable people and regions affected by climate change. There are a number of adaptation funding organizations and some have been mentioned throughout this book. One such organization, the Global Center on Adaptation (GCA), founded in 2018 and based in the Netherlands, believes that 'mobilizing US$25 billion to scale up innovation and transformative actions on climate change adaptation across

Africa' by 2025 will have a significant effect in the forthcoming years. At a recent gathering of African leaders in 2022, half of US$25 billion (US$12 billion) was pledged, placing pressure on rich Western nations who are most responsible for climate change to follow suit. It has been calculated that Africa will need an annual investment of $US120 billion for climate adaptation, but where only $US5 billion was committed at the last COP27 conference in Sharm El Sheikh, Egypt. One of their main programmes is the African Acceleration Adaptation Program aimed at accessing 'climate-smart digital technologies, and associated data-driven agricultural and financial services, for at least 30 million farmers in Africa, supporting food security in 26 African countries, and reducing malnutrition for at least 10 million people' and integrating 'several initiatives to build climate-smart agriculture and resilient food systems in the region'. Regions earmarked to receive financial and technical support for adaptation include the 'Program to Build Resilience for Food and Nutritional Security' in the Horn of Africa to cover '1.3 million farmers', the 'Program for Integrated Development and Adaptation to Climate Change in the Zambezi Basin which will lead to 400,000 farmers adopting climate-smart agriculture techniques', and the 'Ethiopia Food Security Resilience Project designed to support 2.4 million farmers adopt resilience-enhancing technologies and practice'.[2] These important initiatives have been a long time coming and particularly for regions such as the Horn of Africa which continue to suffer from years of severe drought, forcing people off their lands and millions at risk of severe food insecurity and famine. While efforts underway to support farmers directly benefitting from climate adaptation programmes appear high in terms of numbers, it is worth noting that there are 500 million small holder farmers across the world. The point being made is not to appear negative, for that does nothing, but to place in perspective the vast scale the world is facing to support and bring help to where it is needed.

Various figures concerning the costs of environmental adaptation investment and the financial returns it will yield have been circulating to raise awareness. One outstanding cost analysis is the global cost of adaptation programmes calculated at US$1.7 trillion a year would yield an eventual return of $US 7.1 trillion to the global economy. How these figures are reached, one supposes, is a collaborative effort between economists, environmental scientists, and specialists from the World Bank, United Nations, and other institutions such as GCA. As previously cited, the 2022 COP27 conference at Sharm El Sheikh, Egypt, $US5 billion was committed to the global Adaptation Fund, which could be clearly viewed as indicative of the disparity between commitments and requirements. Getting the global community to come together let alone to commit to raising the finances required appears fanciful and even more so given worldwide human conflicts, inflation levels, and countries' individual pursuit of economic growth at any cost. Another issue facing adaptation programmes is the ability of most at-risk nations affected by climate change to secure funds on the global financial market. In their attempts to build climate-resistant infrastructure, developing nations are subjected to paying far higher interest rates on borrowed money than developed Western nations for they are seen as 'at-risk' investments by the global financial market. But when these constraints, bias, and opposing values to securing finances to combat climate change are considered alongside global military expenditure, which in 2022 exceeded $US2 trillion, the values of humankind are its worst enemy.

Once unleashed into the atmosphere, pollution knows no boundaries and the big greenhouse gas emitting countries are beginning to own up to the global impact of their rampant use of resources, manufacturing, insistent growth, wealth gap, and environmental devastation caused across the world. Once stable ecosystems now struggle due to the effects of climate change and become dead zones; lost to acidification, desertification, extreme heat, and depleted fresh water supplies. In Africa, the livelihoods of tens of millions of people hang in the balance, and worldwide 3.4 billion people are at risk due to climate change and 2 billion people live daily with food and fresh water insecurity. The job of various global climate change adaptation programmes is to ensure that financial commitments to support new agricultural practices and deployment of new technologies and infrastructure building are going to those communities most in need so that people can restore and manage their degraded lands. At present, the global Adaptation Fund agreed at COP27 conference is nowhere near reaching its target (at the time of writing, it stood at 20%). Responding to climate change requires active adaptive resistance to the rising temperature of the earth's surface. There is a belief that climate change will naturally correct itself and if Carl Sagan were still alive, he might remind us that there is no such thing as belief, only fact. As the aforementioned summary shows, in terms of climate change policies and the pursuit of universal sustainability, humanity moves in rotating oscillations between reality, effect, distortion, and fact in a pivoting spiral of upward and downward movements of success and failure.

In *The Theory of the Earth*, Thomas Nail, as noted in Chapter 4, charts the formation of the planet from within the earth's magma to its outer surface crust, suggesting that the earth, as matter, is continually in movement. 'Not only is the earth outside itself because of exo-planetary meteoroids, it is also outside itself because its outside crust is only a temporary externalization of a deeper internal process. The earth is a Möbius strip',[3] he contends. We know the earth spins on its axis and rotates around the sun, but via the Möbius strip analogy Nail refers to the earth's inner and outer movement as being spiral. Also cited in the same chapter, Joanna Bourke takes the view that 'the image of the Möbius strip provides a way of challenging tyrannical dichotomies such as biology/culture, animal/human, colonizer/ed, and fe/male'.[4] Timothy Morton in his book, *Being Ecological*, describes the Möbius strip as a minimal topology that 'has no starting or ending point' for it is 'a surface that veers all over, where a twist is everywhere. Appearance is the intrinsic twist in being'.[5] Utilizing the analogy of the Möbius strip with Earth and ground movement can illustrate the mobility of atmospheric pollution and weather turbulence to human adaptation and mobility across the earth's surface as being one and the same with climate change. The driving forces behind these movements—greenhouse gas emissions, atmospheric pollution, ground movement, and human mobility—are not treated as resulting from the same systemic condition. In other words, human mobility is neither free nor equal to the production and oscillation of pollution. The composition of pollution in the earth's atmosphere is finite in terms of measuring the micrograms of fine-particle concentration in the air (PM 2.5) as it is infinite in terms of density and spread. Both determinate and indeterminate concentrations of CO_2 veer all over the atmosphere in a Möbius strip of oscillating weather turbulence onto the surface of the earth.

As the analogy of the Möbius strip suggests, the effects of climate change are not linear; they spiral up and down between the atmosphere and the earth's surface. Encapsulated on a single spiral surface, climate change moves on turbulent weather patterns with neither beginning nor end. As a result, the effects of climate change become difficult to contain, being unpredictable and continually in movement above the surface of the earth. Turbulent weather events are now both inside and outside of human comprehension, both within and yet beyond the scope of predictive weather management. In his book, *Topologies of Power: Beyond Territory and Networks*, John Allen also picks up on the analogy of the Möbius strip, writing that 'when stretched along its length, [it] reveals through a process of twisting locations on opposite sides that are in fact part of the same relational arrangement'.[6] In all these mentions of the Möbius strip there is the same-sidedness of relational connectivity. All life on the earth sits on the same twisting surface, and the building blocks of climate change adaptation are escapable. As the climate change spiral has neither beginning, middle, nor end, and atmospheric pollution disregards continental boundaries and nation-state territorial borders, it has managed to split nature's universal interconnective system and yet unite climatic turbulence in tandem with the mobility of climate change refugees. Prone to spread ever wider, the spiral connecting climate change, oscillating atmospheric pollution, and human mobility is inseparable. Yet, as is obvious, in many wealthy nation-states, this very inseparability is denied, as are the human rights of climate change victims for self-determination and survival, which are restricted if not rejected.

The rights of people living in regions most vulnerable to climate change have been compromised by rich developed and developing countries' CO_2 emissions in the earth's atmosphere thousands of kilometres away. Groups of people are now addressing the immorality that has come on the back of climate change—a recent example can be found in the 2,000 women who took the Swiss government to the European Court of Human Rights (ECtHR), claiming that climate change directly impacts on their lives. Submitted on 29 March 2023, this lawsuit is interesting for the fact that these women come from one of the world's wealthiest countries and are able to access the ECtHR, while tens of millions of people far more affected by climate change have no access and no rights, a further illustration of the great disparity of human rights that exists between rich and poor, specifically, the so-called Global North and Global South. The Swiss women's lawsuit is to be welcomed for it contains the right to life and health for all citizens and calls for the Swiss government to do more to fight and protect the lives of its people.[7] Their view is that if climate change is allowed to spiral out of control, then so too will their health, and as such they are simply protecting their human right to secure their health.

If we are to agree that CO_2 emissions move across the earth's atmosphere and surface in a twisting spiral of connectedness, then whatever happens on one part of the spiral is connected to every other part of the spiral. Thus, climate change producers will inherently become climate change victims. At present the level of threat climate change poses to the survival of people in rich Western nations is not at the same level as those people in poorer and undeveloped nations—e.g. the Horn of Africa and North Africa including Niger, Malawi, and Chad; and Asia such as Bangladesh, India, and Pakistan; and Pacific Island Nations such as Fiji and Vanuatu; and the Indian Ocean such as the Maldives. The Möbius strip analogy is used to tie the world together on the same twisting surface of climate change where humans and all living things on the earth are intricately connected. Human impact is sending shockwaves throughout the earth to a far greater conflict than anything humans have experienced. Humanity, pollution, and environmental devastation now oscillate as one interconnected collision that will drive the future outcome of all life on the earth.

To avert further fracturing of the earth's ecology demands more responsive action and a rethinking of human habitation. To illustrate how such a rethinking can be visualized, a second analogy will be useful: the concept of swarming. In *Facing the Planetary: Entangled Humanism and the Politics of Swarming*, William Connolly argues that societal structures from governments to institutions at local, national, and international levels are augmented through the politics of swarming. 'The politics of swarming, then, is composed of multiple constituencies, regions, levels, processes of communication, and modes of action, each carrying some potential to augment and intensify the others with which it becomes associated'.[8] To associate the complexities of swarming to the effects of climate change born of volatile collisions of atmospheric temperatures on land and sea is to formulate their interconnective yet unpredictable movements in all directions. Matching a turbulent swarming atmosphere to a swarming mobile Earth population falling victim to changing conditions indicates a new human habitation across the earth from the sedentary to the mobile. Swarming does not adhere to a hierarchy of direction but rather movements based on individual connectedness. For example, to witness a swarming flock of sparrows in motion is to see the complex intricacy of mass movement of individuals operating through interconnected spatial awareness between them. As catastrophic weather events are increasingly difficult to predict, human responses must be exceptionally intricate and flexible, embracing all possible directions to minimize the effects. The model of swarming can associate people and climate change in a collective mobility of cause and effect to redirect its impact by spreading individual autonomy alongside collective strategy.

The history of human collectiveness has, on the whole, been elusive where the assembly of complex interrelations between individuals have not come from a basis of equal intention. The ancient Greek *agora* modelled citizenship on inclusive democratic representation of the individual yet it was subservient to the authority's decrees. Both democratic representation and authoritative decree created an ideological indoctrination of the importance of the individual framed through the exploitation of nationalism, patriotism, gender bias, and religion to advance society. The ideology and movement of swarming on the other hand formulates its own model, one that is constantly changing in real time with no certain direction or unifying force. 'The more compelling thing is to gauge the severity of the overriding issues, the time dimension on which they are set', Connolly assures us, allowing for 'the possibility of an emergent, swarming movement'.[9] To engineer a society along the same lines in response to catastrophic weather events would amount to a model of constantly emergent societies, forming and reforming, flexible and indefinable in living with climate change. As with a swarm of sparrows, in which looping movements utilize spatial dimensions in x, y, and z coordinates shaping both mass and direction in real time, so too might new human societies be massed and shaped in connection to climate change. Each sparrow senses the spatial position of another and that spatial awareness radiates across the whole flock. Furthermore, the speed of movement is individually negotiated between each bird and at the same time collectively determined, forming a network of devised directions to shape their swarming. As such no single sparrow dominates another in shaping the collective form, and no shape is repeated; each is adaptive and responsive as both group and individuals. The complexity of this collective moving image is a complex theory that can be translated onto humans to creating new societies living with climate change. Reflecting climate turbulence as a means of response to the effects of climate change ultimately increases forms of adaptation. Present programmes aimed at tackling climate change remain largely untested at the scale required to halt global warming. In other words, human adaptation is not set to a raft of reflex responses that presently characterize global sustainability policies but human controlled responses that fold-in and evolve-out of climate change. As we understand climate change is the movement of altered atmospheric forces and its effects on the surface of the earth, the analogies of swarming and the Möbius strip give a figurative image of how humanity can respond through both collective and individual self-determination, wherein each individual and group adapts and reforms their societies outside of the enclosed systems of government and capital. The swarming model formulates the spatial autonomy for human mobility across the earth as a possible future direction for humans to live with climate change. Swarming involves building a diversity of human responses to the catastrophic problems of climate insecurity and shaping human habitation allied to global sustainability in the 21st century.

The history of early human migration across the earth's continents created the diversity of cultures and languages that would later form nations and countries. As nation-state relations ebb and flow to geopolitical disagreements and conflict at the macro-scale, another level of human cooperation at a micro-scale needs to develop in response. Global policies to tackle climate change can be disrupted at any moment—the Russian war in Ukraine, regional conflicts, ongoing illegal logging, and mining are recent examples of humans' inability to combine at a global level. Then there are the existing disruptions of resistance: the physical and racial borders that divide humanity and nations through the restriction of people's capacity to seek new avenues for their survival. Global sustainable policies are not sustainable if they do not support inclusive mobility, multivariant adaptations, and new societal structures.

To formulate new human patterns of collective swarming mobility in response to weather turbulence is not with its problems, particularly in light of the speed with which catastrophic climate change events are unfolding. 'One problem with a multiregional, pluralizing, swarming strategy that crystallizes in cross-regional strikes is that the movement may take too long to mobilize as the irreversible effects of the Anthropocene accelerate', Connolly notes. 'This too constitutes a live dilemma. In a world of tragic possibility there is no guarantee that the need to act will be matched in fact by timely action. If and when such a mobilization crystallizes, uncertainty will also be high'.[10] Looking at the history of humanity's ability to delay intervention on issues concerning its providence, which has often ended in tragedy, this might be put down to a lack of collective will. By the fact of its autonomous nature, proposing a human swarming response to climate change while creating uncertainty of direction crystallizes collectiveness and innovation for new forms of human behaviour.[11] The multiplicity of spatial orientation that a swarm of birds undertakes is unconsciously transmitted through proprioception—a spatial sensing between themselves, the space that separates, and the space that surrounds them. To differing degrees, all living things on the earth have inherent proprioception abilities. One positive effect of climate change is to bring humanity back to a proprioceptive relation to the natural world. The following section explores adaptation and the sensorial as the primal links to humanity's immaterial self.

Reforming—Living-with, Building-out Nature

Building the physical environment of cities and associated infrastructure and technology removed the existing natural environment to assist speed and mass urban living. Parcelled into neighbourhoods separated by degrees of wealth and privilege, the city became an all-in-one habitat servicing its occupants' every need. Nature in cities is mostly non-existent, for city dwellers have little need of it and where it does exist, like in parks and gardens, it is human-made. Living without nature and losing connectivity with natural ground laid over with artificial surfaces, cities have become islands of dislocation from that which lies beyond its boundaries. Technological and intensive consumption coupled with the ability to restrict and control heat, cold, wind, and rain behind sheets of concrete and glass have modified human behaviour to live separately from the natural elements. The devolution of nature from human experience in cities formulates a barrier to the natural world and also limits people's understanding of the real effects of climate change in those regions where it is felt most. The separation between city dwellers' and rural populations' experiences of climate change is a course for concern to achieving sustainability. Where Chapter 4 explored animal and plant life to find how humanity can learn from their ecologies to realize closer relational partnerships with the natural world, this section explores how human behaviour might be reformed to a naturally occurring relationship in response to a rapidly deteriorating Earth.

In *The Dominion of the Dead*, Robert Pogue Harrison suggests that the earth's demise in the human experience is the result of ceding to a new conception of the earth. Harrison does not directly refer to the earth's demise in relation to the conception and construction of the city but rather its propensity for environmental destruction to achieve it.

> A truly extreme or self-consuming vision of annihilation takes the form of the earth's demise, for the forces of destruction, when pushed to their comic extreme in the human imagination, not only destroy all that human labor builds in time, they also destroy the supporting element of time, namely the land on which we erect our worlds.[12]

Cities altered not only human contact with natural ground but also their ancestral associations. First nations peoples kept their ancestral connections and claims to the land on which they reside through the burial of their ancestors, geographical knowledge, and customs. In another book by Harrison, *Forests: In the Shadow of Civilization*, he ties ancestral burial and ground, whereby '[b]urial guaranteed the full appropriation of the ground and its ultimate sacralisation. Through burial of the dead the family defined the boundary of its place of belonging, rooting itself quite literally in the soil, or humus, where ancestral fathers lived underground'.[13] Through the dead, ancestral burial sites were a prime reason for demarcating ground and ownership. Australia's first nations people refer to their homelands as *Country* as an extension of themselves and the sacred burial sites of their dead. Country embodies the spirit of belonging and communication to the land. Early anthropological thought often confused native people's embodiment of nature—geography, water, animals, plants, spirit etc.—as inferior and naïve, at odds with a 'superior' Western culture where nature exists for humans' benefit. The flawed analysis in Western anthropology to first nations peoples worldwide is crucial for understanding present-day climate change wherein Western dominance has prevailed through the destruction of first nations peoples' culture and environment.

Since the transference from nomadic life to permanent settlement, human existence has depended on altering its surroundings, cultivating crops, and consuming animals for survival. Nature also engages in the same process, following the path of natural selection where animals and plants maintain a balance in the diversity of species, predator, and prey. The two dominant types of human existence that have emerged over time are evenly split between the world's four billion people living in dense urban centres and the other four billion living in rural regions. As pointed out, the distinction between the two contains the erasure and exposure to nature. Given their different dependencies on nature, urban and rural populations are nevertheless joined by their ability to shape the ground: one by human-made separation of nature; the other through the cultivation of ground with nature. Through different devices, such as the plough to furrow the soil to tools and materials to construct the city, rural and urban populations exert control through exclusion and inclusion of their surroundings. The more successful humans were in constructing and cultivating their environment, the more they erased the origins of the nature surrounding them.

In *Designing the Earth: The Human Impulse to Shape Nature*, David Bourdon highlights how the cultural diversity of societies shaped their relationships to geography, landscape, and ground. He examines shelter, burial mounds, war defences, vast mining works, and land art through geological terrain shaping. As human existence is based on resourcing, the earth and its capacity to conceive, design, and manufacture mass products has come at a huge cost to plant, animal, and ocean ecologies. Excessive resource extraction and over-farming has turned once fertile lands barren, contaminated rivers, and devastated forests and oceans. It has done the same to the atmosphere, turning what was once invisible air into visible toxic pollution. As it stands now, to sustain human progress is to continue to keep extracting resources, threatening plant and animal life in return for humanity's ongoing existence. With control over the natural environment came superiority to shape the human psyche; the more successful they were in shaping the earth, the less need there was to maintain its visible origins. 'Power is at the core of the geo-constructivist conception of the world. The power of humanity as a kind to reduce its status of a species to its smallest part; a power that takes place on a devitalized planet', so Frédéric Neyrat argues in his book, *The Unconstructable Earth: An Ecology of Separation*. Utilizing the 'technological dimension', 'geo-constructivist power', and 'technopolitics' enabled humankind 'to rebuild the planet according to its own desires' under the umbrella of 'terraforming'.[14] The sense of urgency to break the cycle of destruction in favour of sustainable cultivation of the earth relies on limiting human desire for unlimited consumption.

Other than 'terraforming' their surroundings, humans have been 'terraforming' themselves. Destroying opposing settlements, raising citadels to the ground, and casting first nations peoples into poverty through invasion and discrimination—these are just some of the ways in which humans have managed relations between them. With conflict and destruction come resolution and reconstruction where the origins of the destructive event are erased. The mass destruction of German cities during WWII and their subsequent reconstruction after the war removed the physical scars of destruction to a form of forgetting. In *On the Natural History of Destruction*, W. G. Sebald refers to a collective forgetting by German society as it emerged from the devastation of its towns and cities. He attributed this forgetting to a type of amnesia, a self-anaesthesia 'shown by a community that seemed to have emerged from a war of annihilation without any signs of psychological impairment'.[15] Sebald's self-anaesthesia of forgetting the destructive event by reconstructing the destroyed city has some parallels with humans overlooking the destruction of the earth's ecologies to preserve their existence. The mentality that persists in the human psyche for the annihilation of cities in war to the annihilation of nature is the belief that rebuilding cities can be applied to restoring nature. What is obvious in the material reconstruction of brick, concrete, stone, and glass in cities is not the same as restoring the natural environment, for the latter lives and dies according to natural laws that govern life. When those laws are broken by human impact, the mechanisms embedded in nature to regrow and adapt diminish.

Human progress has come at the cost of toxic air, polluted oceans, depleted glaciers, degraded land, deforestation, and extinction of animal and plant species. The vast construction of mega cities, infrastructure stretching across landscapes, mining, and the manufacturing of billions of products have turned the earth into a service industry for humankind. To restore the earth, humanity must shift from its 'natural' propensity for destruction to a shared common existence. Nature cultivates its life force; it is sentient and holds the memory of the earth but when that life force is depleted or erased, its ability to regenerate becomes irretrievable. To adapt to climate change, humanity is deploying the same methods it applied to removing and processing the earth's natural resources by reengineering the same technologies. It is not humans who are adapting to reduce their impact on the earth, rather humans are trying to 'hack' or 'tech' the natural environment without jeopardizing modern living standards. Vast resource extraction, manufacturing, and consumption is still chosen as the model for humanity to live on the earth, which belies a 'ferocious fantasy: recreating the Earth, reconstructing life on Earth—to the point of repudiating death', in Neyrat's words. Humanity's belief in itself is such that the act of continuing to destroy the planet is equated with its ability to restore it; yet, 'as long as the impoverishment of biodiversity will not have cruelly devastated the planet, humans will continue to *grant themselves an ungraspable kind*: Humanity is only a species when it discovers—after the fact, after every ecological disaster—that it lacks power'.[16] The reality in which humankind is prepared to believe, accepting environmental destruction as part of human life on the earth, betrays its crazed sensibility. To stare at the brink of environmental collapse before it grasps what to do is a sobering indictment of humanity's psychological condition and relationship to all other living things.

For decades now, humanity has possessed the technical ability to destroy life on the earth through nuclear warfare. Besides the initial mass destruction of cities and people is the enormity of nuclear fallout across the world. Thickened by clouds of dust covering the sky where the sun no longer penetrates, deadened landscapes where nothing grows, humanity is left with the death of itself. We have seen enough documentaries and science fiction films to know what the outcome of a nuclear conflict brings.[17] As humanity is already living in its own creation of environmental destruction that will shape the earth's future for millennia to come, the possibility of nuclear war on top of environmental collapse speaks again of an unstable human psyche. As more countries gain nuclear-strike capabilities and the scaling-up of a new nuclear arms race grows, the incomprehensible becomes comprehensible.

In the light of humankind's capacity for destruction, the question is: how can humanity allow nature to re-emerge? Humans are starting to come to terms with their impact on Earth as manifested in catastrophic weather events, each more devastating than the last. As humanity's denaturing of the earth and the results of climate change become more apparent, it has questioned its connections to the natural world. Humans are now attempting to unearth their buried connections to ground and environment by seeking to re-engage the dormant 'nature' within their bodies. Epigenetics—an area of molecular biological study focusing on the changes of organisms caused by modification of gene expression rather than the alteration of the gene code itself— offers a solution (if a potentially disturbing one) to programming changes in human adaptation to radically changing ecologies. In the course of human progress, the human genetic code has undergone modification, transforming humans' relationship to the natural environment. The long-term effects of climate change are forcing humanity to act and redefine its 'nature' to the natural world. The present programme that humanity is pursuing is the creation of agreements and targets between countries to reduce their carbon footprint. First came the policy of a global trading scheme in CO_2 emissions, which companies and countries established in 1997. This proved to be detrimental to tackling climate change for it bargained the responsibility of polluting industries and fossil fuel companies to reduce their CO_2 emissions to a marketplace stock exchange. The scheme was essentially a *faux* eco-caring game that allowed fossil fuel and polluting industries to continue with business as usual by buying and selling carbon credits between them.

The overwhelming scale and mystery of the natural world exerted both a familiar and haunting experience on early nomadic life and settlement. To recall Joanna Bourke's argument, it was men who brought fear and timidity into the world and subjected it onto women to construct a form of social control and conformity in their own image. To further appease their self-constructed trauma, men turned their attention to shaping the earth, bringing a sense of control over nature and society. In *Haunting Legacies: Violent Histories and Transgenerational Trauma*, Gabriele Schwab suggests that trauma 'as a mode of being violently halts the flow of time, factures the self, and punctures memory and language'. Schwab argues that '[s]ome lives are hit with catastrophic trauma over and over again; then trauma, with its concomitant strategies of survival, becomes a chronic condition. Defenses and denial become second nature; traumatic repetition becomes second nature'.[18] The natural world was not an enemy *per se* for humanity but a problem in itself. Working through the haunting legacy of fear and timidity, men cemented their ideology through patriarchy to dominate society and extended it to shape and dominate nature. For humanity to succeed in over-coming its haunted legacy of 'natural destruction' and create a viable sus-tainable future to tackle climate change in the 21st century, it will require a restructuring of the patriarchal system as well as removing global inequality and environmental devastation.

The present approach and implementation of sustainable policies are geared to restoring devastated environments by applying an exterior layer between humans and the natural world. Sustainability has become an outside operation to counter the catastrophic effects of climate change rather than an absorbing interior programme of human adaptation with animal and plant life. The male construction of fear and timidity projected onto women and the environment has re-emerged from the effects of climate change, constructing a new layer of fear and projecting onto an unpredictable nature. Adapting to the world's radically changing ecologies involves transforming human behaviour and removing its separation from the natural world, animal, and plant life. In her chapter, 'Fluid Bodies, Managed Nature', Emily Martin refers to the problems associated with separating humans from nature, which overlook how 'bodies that flow easily into spaces beyond the skin create the potential for disturbingly labile forms of association; these associ-ations, whatever their merits, are unmoderated by the brakes of categorically opposed divisions between the human and natural worlds'.[19] To remove the new fear of living under climate change is to reignite human embodied nature as an ecological wearable skin of environmental restoration. The chance for a lived ecology free from the haunting legacies of environmental devastation that have shaped human interactions with the natural world is part of its shift to achieving global sustainability.

In the 1960s and 1970s, radical art and social movements sought to present versions of a future world where technology, humanity, and nature converge as compatible and relatable where each is a subsid-iary of the other. Visionary future worlds cross-pollinated art, architec-ture, and ideology to forge an integrated technological formalization of nature in their projects. Groups such as the Italian futurists, Superstudio and Archizoom and United Kingdom's Archigram, French spatial/urban theorist, Guy Debord's Situationist movement of urban drift, the *dérive* later championed by Dutch artist, Constant Nieuwenhuys in his project 'New Babylon' are just some examples of an open urban and geographical movement between the people, the built environment, and nature. Dissected urban plans, super-structured architectures traversing across cities and geographical expanses, their collective ideas were to conceive spatial transformations for new inhabitable Earth futures. More vanity projects than realizable schemes, their superimposed conductor networks of superhighway infrastructures straddled cross-cultural, transitional, and continental landscapes as a form of resistance to the archaic social structures that dominated human life in the latter half of the 20th century. Collectively their projects were not contextualized to fit a certain site, city, region, or nature but were un-sited and applicable to a generic conception of human habitation across geographies. Straight lines, grids, and smooth surfaces dominated their architectural vision in contrast to organic forms of nature. Depending on the viewers' imagination, their schemes could be as real or unreal in deciphering the graphics, collages, and models these futurist groups produced. Their work was a celebration of human achievement in bending societal visions, and even though their projects did not necessitate finding a balance with nature, their work nevertheless broke the conventions of stability and permanence in favour of a mobile band of global, self-determining inhabitants.

The works of Debord, Constant, Superstudio, Archizoom, and Archigram and others have not been lost in the 21st century. In terms of scale and imagination, no more is their vision of a superhighway–superstructure society more apparent than in Saudi Arabia's NEOM Project. At 170 kilometres long and 500 metres high, this vast one trillion-dollar-plus housing and commercial city dissects its desert landscape in a straight line from the sea to the mountains. Designed by the American architecture studio Morphosis, the scheme is to house and provide employment for seven million people. The design consists of two glazed buildings stretching the distance that deflect as much as reflect the harsh climatic conditions of the Arabian desert, while between them a garden paradise flows in a constructed topography bringing natural cooling and human wellbeing. In the advertising materials for the project, the narrative spun is as convincing as it is disturbing to its many critics.

> For too long humanity has existed within dysfunctional and polluted cities that ignore nature. Now a revolution in civilization is taking place. Imagine a traditional city and consolidating its footprint, designing to protect and enhance nature. The line will be home to nine million residents and will be built with a footprint of just 34 km². And we are designing it to provide a healthier more sustainable quality of life...Residents have access

Figure 6.1 Line city excavation work for NEOM Project, Saudi Arabia. Aerial drone image taken on 13 October 2022.

Source: Image courtesy of Otskydrone.

to all their daily needs within a five-minute walk and the line's infrastructure makes it possible to travel end to end in 20 minutes with no need for cars.[20]

Trumpeted as a sustainable project whereby its linearity reduces its carbon footprint compared with that of a traditional radiating city, the monolithic NEOM Project is beginning to take shape with massive earthworks underway. The project has many detractors in the fields of architecture, urban design, and environmental science, including writer and urbanist Adam Greenfield who describes the project as an 'ecological and moral atrocity'.[21] If the project's rendered images are to be believed, they constitute a fusion of reality and fantasy to 'see what happens'. The excavations presently underway are vainer and more exclusionary to the natural environment than any of the projects Superstudio, Archizoom, Archigram, Debord, and Constant conceived. The NEOM Project, where unlimited finances, construction, and technology can realize a city in the desert for nine million people and this can somehow be described as 'sustainable', tells of an approach to building-out the environment as much as building-in climate change. The carbon footprint cost of transporting materials and construction on such a vast scale and in such a harsh environment appears on every level opposed to sustainable design and building. Its programme is untested as its environmental damage yet unknown. The NEOM Project propagates the superior relationship humans have constructed with the natural world by design and by conditioning the environment to be 'fit for purpose'. There is no reforming of the relations between human technological prowess and the environment in this project. In fact, it takes human society and inhabitation on the earth to the extreme of separation where neither connects with the other.

Before mass human impact on the earth, the earth evolved from natural collisions forging an array of diverse ecological conditions. The earth's environmental self-determining diversity has now been ruptured, creating a tipping point in both atmospheric turbulence and surface devastation. To counter the threat posed to human existence by an unpredictable, turbulent Earth requires a reforming of human superiority over nature. No building, sustainable policy, or ideology can elevate nature other than nature itself. The course to be undertaken by humanity is to rediscover its 'nature' to transform and create new variations of human life on the earth. It is a course to be taken individually and collectively, without government interference, and, if necessary, by force. It must be radical, planned as much as felt and sensed in building inclusion, cohesion, and interdependence between humanity and the earth's ecologies.

Variations—Biological Departures

The population of displaced people around the world is set to double within the next 20 years. The United Nations conservatively calculates there are 100 million people displaced around the world, while other organizations place the number at closer to 300 million. With two billion people facing food and water insecurity daily and with catastrophic weather events due to climate change set to become more devastating, the challenges will stretch the available finances and capacities of countries, global institutions, and NGOs. What has become clear is that human impact on the earth's atmosphere, land, and oceans is creating a mass movement of displaced people who are either temporarily housed in refugee camps, incarcerated in detention centres or living in makeshift shelters at border crossings. This is fast becoming a permanent condition of life in many parts of the world. In *Drifting: Architecture and Migrancy*, Stephen Cairns describes 'refugee camps' and 'detention centres' as facilities 'designed to control and deter unauthorized travel of refugees and asylum seekers across national borders'.[22] With a growing population of displaced people, a new set of terms will be needed to identity them. Leaving their homelands due to environmental damage such as desertification and drought brings the 'illegal' status concerning their mobility. Whether they are identified as climate change refugees or economic migrants, both terms stigmatize their plight. Whatever routes displaced people take, their mobility is a threat to countries they seek to enter to better their chances of survival and the only way to stop them is to incarcerate or repel them. Climate change has created a new level of global discrimination on top of already entrenched racial, religious, and cultural discrimination between people and countries. It is time to rethink who these people are not in terms of fear and threat, race, and culture but as new frontier explorers forcibly made to adapt to climate change with the least of resources available to them.

Cited throughout the book, the territorial sovereignty of national borders become notional when considered against the nonbinding borders of climate change. Territorial borders between countries in Europe for example came into being through culturally binding associations of differences in language, cultural identities, and war, but as pointed out in Chapter 5, territorial divisions were also imposed. During the colonial era, the African continent was carved up by its imperial rulers along lines that cut across cultures and languages leading to 'made-up' countries of non-associated clans of people. It has taken post-colonial countries generations to reform their cultures—much of it still ongoing and some of it irreparable. As the terrestrial movements of millions of displaced people around the world grow, so too will new geopolitical and anarchial-geographical forms of resistance to borders of separation. Simon Springer in his article, 'Anarchism and Geography: A Brief Genealogy of Anarchist Geographies' calls for the contestation of people, place, and ground to claim their rightful passage of existence where 'environmental justice and sustainability' go hand in hand. He identifies an 'anarcho-geographical perspective' citing a range of phenomena: 'informal economy, livelihoods, and vulnerability; cultural imperialism and identity politics; biopolitics and governmentality; postcolonial and post development geographies; situated knowledges and alternative epistemologies; and the manifold implications of society–space relations'.[23] Connecting Springer's list of historical and contemporary human injustices to globally displaced people as a result of climate change reveals the challenges as much as the potential for new societies to be born in terms of terrestrial mobility and spatial occupation. Where displaced people occupy the highly defined spaces of inter-territorial borders separating countries, their identity is cross-bordered and transnational to their bodies, which are hard-pressed to barriers of steel and razor wire. Whether viable or naively speculative, a reworking of territorial borders has to be taken into consideration concerning displaced people and climate change adaptation in the 21st century. When countries, institutions, and NGOs are unable or unwilling to accommodate displaced populations, this means that the only recourse these people have is to empower themselves and create their own pathway however fragile and perilous it may be. We witness the human cost daily in their journeys across deserts and seas and, as history has shown, where discrimination, rejection and oppression exist, so does resistance.

Global ecological disasters are an after-effect of human impact on the earth's ecological system. Under stress and facing inevitable decline, nature's ability to rebound from centuries of human impact is no longer assured. Agreements struck at the 2015 Paris Climate Summit to limit global warming to 1.1 degrees Celsius from pre-industrial levels, reach carbon neutrality by 2050, and build climate change resistance are failing. The latest figures forecast that the earth's temperature will rise beyond 1.5 degrees if not higher by 2050 and a further one billion people will be unable to support their lives in the regions where they live. This could possibly mean one billion displaced people on the move, interred in refugee camps, filling already stressed dense urban centres, accumulating at borders and dying crossing deserts and drowning at sea in trying to reach rich Western nations for a better chance at life. Most of these people, if not all of them, would have had little impact on global warming but have paid the highest price because of it. At present, the global community is unable or unwilling to support programmes for climate change adaptation on the scale required. As previously mentioned, the COP27 Adaptation Fund is nowhere near meeting the financial commitments it has promised, and divisions between the so-called Global North and Global South continue to increase rather than decrease, despite this issue being one of the main focal points of the COP conferences over the last 27 years. The effects of climate change remain grossly inequitable between populations affected (the majority) and populations responsible (a minority) for the bulk of global carbon emissions. The volatility of nature is challenging whole populations, countries, and continents. It is safe to assume that never before in human history has there been a common threat on a global scale and at the same time. Given that humanity has lived constantly under a cloud of human conflict, invasion, and war and as the 20th century proved the ability for mass destruction does not override the fact that more money is spent on military projects that in 1 year alone would finance commitments to fight climate change over the next 10 years. There is a distinct lack of integrity where the human propensity for conflict still outweighs its capacity to resolve opposing ideologies and collectively assemble to fight climate change and reduce worldwide poverty, meaning that reaching global sustainability is seen as a mere hope rather than an absolute necessity.

Where the 27 COP conferences on climate change and sustainability have been charged with forging global alliances and collective action on global warming, it is becoming a distinct possibility that restoring the earth's ecological system is now beyond human capabilities. The reality of positioning climate change as the central focus for future human and Earth existence is blunted by the reality of geopolitics, ideological war, and international trade. Returning to Frédéric Neyrat's analysis, to understand humanity's cataclysmic psychological condition where the 'impoverishment of biodiversity' accounts for 'ecological disaster' is to understand how geopolitics and power work across the world. 'Power is at the core of the geo-constructivist conception of the world. The power of humanity as a kind to reduce its status of a species to its smallest part; a power that takes place on a devitalized planet'.[24] Learning how to remain positive and instigate environmental adaptation programmes amid an ideologically divided world is where the disruptive actions by groups such as Extinction Rebellion, Just Stop Oil, and Last Generation step into the picture as a way to strike against local governments and global corporations to force change. The recent arrest and criminalization of Last Generation protesters in Germany and 1,500 Extinction Rebellion protesters in the Netherlands tell of the immense failure of authorities to grasp reality. Their message is simple: humanity's technological progress over the last two centuries sowed its sense of invincibility to the forces of nature and now, in the 21st century, nature is sowing human vulnerability such that something has to be done.

How might humanity change its perception of nature from being solely a provider of resources to being part of a cooperative, inseparable, and protective relationship? Is it possible for humanity to change the direction it has taken over thousands of years of shaping the natural world to service its needs, knowing that to continue threatens its very existence? Answers to these questions would not need to be considered if humanity embarked on a path by considering nature's existence as inseparable from its own. '*Nature and humans have never been separate systems*', Thomas Nail says in *Theory of the Earth*, and taking the present epoch of the Anthropocene into account 'is not only about humans and what they have done to the earth. It is about the earth and what it is doing to itself through humans'.[25] Humanity coopted nature for its purposes and now humanity is reconfiguring nature to coopt it. While humans and nature have never been separable entities, humans forged their divergence from the natural system of the earth. 'We either act as if our scientific knowledge about the earth is a separate thing, unconditioned by the earth itself, or we think that the earth that existed before us and will exist after us is somehow radically unrelated to us'.[26]

Nail views humanity's single-minded perception of the earth as dismissing its geological, centripetal, centrifugal, tensional, and elastic formations.

> By thinking only about our own movements of energy expenditure and conservation on a 'relatively static earth,' we have failed to see ourselves as part of the larger cosmic and terrestrial drama of increasing flow rate and mobility. By damaging the earth's dissipative processes (especially the biosphere), humans have slowed down the kinetic movement of energy throughout the planet. Fossil fuel capitalism has increased human energy consumption, but only at the cost of decreasing planetary energy consumption by much more.[27]

The disruptive nature humans exerted on Earth has resulted in dissipating nature's kinetic interrelations, severing material from non-material matter. Human dominance over the earth clearly hedged its ability to adapt to the earth's environments and aided by the industrial and technological revolutions, humanity evolved not into a new species such as happens in animal and plant life but adopted variants.

As cited earlier in this chapter, the science of epigenetics allows biological modification without altering DNA sequencing in order to molecularly restructure adaptative evolution of animal and plant life. In *Stages of Transmutation*, Tom Idema writes of his interest in human transmutation, which lies in the written, visual, and technological worlds of literature, science fiction, and environmental posthumanism. He notes how the science of '[e]pigenetic transmutation allows for a quick adaptation to an overwhelming modern world'.[28] Biological human re-engineering through epigenetics was made possible following the DNA sequencing of the human body known as the Human Genome Project. Resequencing human DNA through epigenetics provides the possibility for biological–transhuman variants to relink humans to the natural environment through cellular biological connectivity. The ethics and moral currency of such a programme are fraught and possibly horrific. As the apex predator of nature, the divide-and-rule mentality humans have deployed until now has come at the cost of losing its affordance. While industrial and technological progress has evolved human life on the earth, it has created itself as victim.

Substituting their evolutionary movement in sequence with the natural world for dominance over the nature, humans limited their capacity for environmental variation and intrinsic knowledges of the land as practised by early nomadic and first nations peoples. Human biological development over hundreds of thousands of years spread the racial, cultural, and linguistic diversity across the continents. Human investment in industrial and technological innovation over the last two centuries tended to flatten-out the human biological cycle and mobility in step with the natural world. 'Kinetic cycles', Nail writes, 'are the metastable basis from which growth, dissipation, multiplication, and so on all move and develop. Development and variation all assume the emergence of a prior metabolic center of orientation from which anything can develop'.[29] Variation is derived from kinetic energy of the earth creating ecological diversity that in turn forged the transmutation of all living matter. The 99% of the atoms that make up the human body, hydrogen, oxygen, carbon, nitrogen, calcium, and phosphorus, tell us that on a micro-biological level humans are essentially no different from animal and plant life or the visible and invisible matter of air, water, soil, and stone. 'Atoms and celestial spheres, the spread of forest and the retreat of glaciers, the flow of magma and the brusque solidity of stone produce an improvised choreography… And yet both stones and humans have their irreducible particularities', writes Jeffrey Cohen in *Stone: An Ecology of the Inhuman*. The 'irreducible particularities' point to the challenges facing humans to transform and adapt not to a simpler condition from where they are now but a more complex, natural edification in the face of extreme climate disruption. The simplified statis of stone belies the complexity of its adaption and movement over billions of years to its surroundings to hold its climate internally as much as being shaped by it externally. Humanity has built and managed a complex nature for its existence by simple means of rational exploitation of the natural world. For humans, the natural world has always been problematic and so the objective has always been to get rid of the problem. 'Nature is a force to harness or subdue, a resource for human culture, a contradiction-ridden troublemaker, and a stuttering promise of human superiority over animals, plants, and minerals', Cohen explains. 'Nature is a source of beauty' he continues it is 'theology, teleology, taxonomy, and physics as well as anxiety, gender trouble, animal trouble, life trouble. Nature is, in other words, a ceaseless problem'.[30]

The departure for new human variation to build a kinetic mobility with the earth and all living things is fraught when considering that the multiplicity of animal and plant variation is being depleted at an alarming rate. Any depletion of species in turn depletes human variation for adaptation. The possibility of epigenetic modification for human biological variation in sequence with the kinetic movement with the natural world, when read in light of this book's concept of 'wearing our ecology', suffers immensely when considered alongside species extinction. 'Wearing our ecology' is not an exoskeletal biological costume but an internal biological condition for human adaptation. Sustainability is not solely the adaptation and restoration of devastated regions, environments, and ecologies from climate change, it is the capacity for human adaptation to radically changing conditions. If humanity only yields to its inventions for technology-based solutions to climate change, rather than genetically inspiring ecological connections, then adaptive living will evolve in sync to a radically changing Earth. To create human variation to living on the earth requires the affordance for transmutation of the human species—the ecological interior that involves 'wearing our ecology'.

Forwarding a biological affordance for human adaptation to living with climate change brings with it a great deal of scepticism. Scepticism has long been deployed by various industries, authorities, and lawmakers to acknowledge the threat of climate change on human existence and the urgent need to implement global environmental and sustainability policies. A similar scepticism persisted from scientists, public, and religious groups at the time of Charles Darwin's theory of evolution. Referred to in Chapter 4, Darwin's transmutation of the species to environmental adaptation threatened prevailing creationist religious views and was considered inherently flawed by scientists for it did not explain the evolution of marine life to terrestrial animal species. Darwin's 'survival of the fittest' concept, based on the optimal abilities of animals and plants to coexist in a participatory existence balancing life and death, was seen as an affront to the notion of human superiority, for it suggested that humans are not in fact superior Earth *beings* when considered alongside the adaptive abilities of the natural world. If human evolutionary transmutation was made possible by its departure from the natural environment, then there is the chance to alter this movement in the face of the climate crisis—this is the task humanity faces in the 21st century. To look back on the last 100 years is to acknowledge that human impact on the earth has exceeded the natural world's ability to absorb its effects, creating what geologists refer to as the Anthropocene. Human adaptation in the face of climate change has to relinquish its controls over the natural world and formulate new living habitats under conditions of catastrophic weather turbulence. But this is only possible if global financial systems are restructured, resource plundering reduced, and universal equality between humans and equity with the earth become the central focus to achieving a global sustainable future. As mentioned, taking the combined global budget spent on military armaments would make the above realizable in the decades ahead but humanity's propensity for destruction seems to have overwhelmed its ability for creative adaptation.

In *How Life Learned to Live: Adaptation in Nature*, Helmut Tributsch offers a concise biological explanation of animal and plant life composition in relation to their natural environment. Tributsch studies the 'transformation of sunlight into chemical energy by photosynthesis in plants', through to the adaptation and alterations of blood flow, heat exchange, light and energy transference between animals and their environments. He describes the inner and outer relationships plants and animals have with their environments— responsive, osmotic, absorbing, and reflecting. How to reimagine human connectivity in a time of a rapidly depreciating biosphere tells of the urgency for human transmutation and adaption to climate change. Although Tributsch's research focuses on animals and plant life adaptation, it is relevant to humans in a number of ways, for example, in adjusting to heat, sunlight, and water, but also in terms of how animals and plants can influence human habitation.

> Since life needs air, and a protective shield, it is in theory subject to conditions similar to those that prevail for a photochemical surface reaction. Such a reaction is the process of photosynthesis in green leaves, by which light is transformed into chemical energy. Perhaps then, nature would build cities similar to the submicroscopic thylakoid structures— the power stations of plants, which consist of self-contained flat membrane sacs, often stacked like rolls of coins and linked to each other by many cross-connections. The units are arranged so as to make maximal use of light and to form as large a contact surface as possible with the environment–architectonic criteria our cities still fail to meet adequately.[31]

One of the steps towards reimagining human adaptation to the natural environment is to re-enter the ecological cycle of life, death, and regeneration— that is, to abide by the laws of nature. What becomes complex in realizing the above, however naïve it might sound, is altering the human system where one faction accepts its dominance over another. As cited in the opening chapter, just 1.1% of the world's population own 45.8% of global wealth; 11.1% own 30.1%, 32.8% own 13.7%, and 55% of the world's population own

1.3%.[32] Issues to address inequality in sharing the world's resources and wealth, reducing fossil fuel use, committing to sustainable living, and advancing global human mobility as part of living with climate change are complex but not out of reach if governments, corporations, and rich Western nations desist from propagating consumption as the key to economic growth and human prosperity. To achieve what many governments and corporations do not want to change may require a people-led global revolution to live outside the authorities' controls and the catchment of consumption. The analogy of the Möbius strip, where every position along its surface is the same as any other, illustrates how non-centralized governance promotes microsocieties generating their ecological coexistence and sustainability across the same surface. Such self-determining societies living freely on the earth and in accordance with their ecological connections with their surroundings are possible; when previously accepted societal structures are no longer viable, they can be taken down and rearranged by the people.

Another way in which human adaptation to climate change can occur is by removing the separation between rural and urban populations and installing new migration patterns between them. As mentioned in Chapter 2, creating greater fluidity between these two distinct halves of the world's population directly connects environmental adaptation and practices of sustainability to shared knowledges and experiences. Imperative to achieving global sustainability, this becomes problematic considering the forecast where by 2050 about 70% of the world's population will reside in urban centres up from the present 55%, which makes the shared knowledges for adaptive living under climate change between rural and urban populations all the more critical. The rapid rise in urban expansion and global population will apply more pressure on urban centres that are already struggling to provide infrastructure such as housing, sanitation, transport, jobs, and energy. The result will lead to substantial increases in informal dwellings, poverty, and homelessness in both developing and developed countries. Existing on the fringes of society, many of these people will, by default, remain outside the framework of global sustainable policies. As the world already exists in a two-tiered 'Global North' and 'Global South', with multiple-tiered economic, education, race, religious, and gender classes, societal, national, and intercontinental separations will continue to widen throughout the world posing major challenges to global sustainability. This global splinter was recognized at the 2022 COP27 resolution to finance adaptation programmes to regions affected by climate change aimed at restoring degraded lands and modernizing farming methods, including increased use of genetically modified drought-resistant crops. But the funding commitments actually provided so far have fallen far short of their promises.

Humanity is placing a great deal of trust in its capacity for technological innovation to solve global warming. The example of carbon-capturing plants currently being built in the United States and Iceland, which are hoped to be able to 'technofix' carbon pollution by sucking it from the atmosphere, is in reality not feasible in light of the scale of the reality, technological machinery, huge expense, and notwithstanding the ability to cover the vast expanses of the planet. On top of the 'technofix' attitude, there is the issue of the timeframe for tackling climate change being set to a future time on largely speculative projects. The goal of tackling climate change cannot be placed in terms of decades such as 2050 and 2080 goals to reach carbon neutrality when carbon emission targets set for 2030 are already unattainable. These goals—both real and imagined—are absorbed in 'tech dystopia' sweeping the world—increased surveillance, dismantling of individualism, consumption fetishization, anxiety and fear of global migration, and environmental depression from many young people, all of which put pressure on global sustainability. Human transformation with the natural environment to live sustainably will come down to individuals fighting against an increasing 'tech dystopian' world. With the establishment of agriculture, humans gained their ability to shape the environment for their own ends and through technology they hope to reform their disruption of the planet's ecological system. Viability, feasibility, actuality, and fantasy are all equally recognizable in tackling climate change through their distortion.

It is clear that the future of human existence on the earth rests on people's ability to reform the entities that polarize the world. In *Creating an Ecological*

Figure 6.2 Climeworks carbon-capturing plant, Orca Plant, Iceland.

Source: Image courtesy of Climeworks.

Society: Toward a Revolutionary Transformation, Fred Magdoff and Chris Williams pursue an ideological route for the establishment of an ecology-based society to counter human impact on the earth. Magdoff and Williams talk of people-structured societies where the drivers of capital, profit, and greed are divested in favour of investment in human ecological living.[33] Their proposal connects to revolutions throughout human history from uprisings to overthrowing the injustices of tyranny and repression, replacing them with a new ideological system based on equity and freedom. Many of these missions failed outright and when they have been partially successful, they have not reached consensus and agreements for collective governance. This often led to factions and fragmentation, leaving the door open to the forces of authoritarianism to regain power. The point is that people tire of ideology—of its ideas as much as its implementation. The same might be said of how the ideology of sustainability—a theory not yet tested globally—fails not only by unproven global implementation but by the ability of people throughout the world to sustain their commitment.

To talk of human adaptation and variation in creating departures for a new coexistence with the natural world is limited when considering the multiplicity of obstacles placed in front of it. To speak of human transformation in adapting to climate change, as raised in this chapter, is not to echo Darwinian theory by suggesting that humans have the same capacity for transmutation as animals and plants—to adapt their form in response to their surroundings. Instead, it is for humans to change in accordance with nature. For sure, the outcome is speculative and there are no guarantees that humans can arrest the extreme weather turbulence from global warming. Human transmutation does not involve a remodelling shaped by its surroundings but the perceptive connectivity to its natural genetics. In this way a radical altering of government and the power of capital in favour of people-led societal orders based on sustainability begins the fundamental shift in human habitation that is imperative to living with climate change. Simple as it is complex, it nevertheless remains unapproachable under the current global system.

Changing the global financial system and people's habitual consumption patterns ingrained with modern living may appear insurmountable. Yet, the evidence of human transformation already underway in people's awareness, eating habits, consumption, and waste management shows that humans are willing to exchange privileges for climate security. On a global scale, humanity is still dealing with age-old issues that have stalled its progress such as violence, terror and war, dictatorship, corruption, and greed and most visibly economic, racial, and gender inequality. This chapter has attempted to present some ideas as to how humans can depart from their historical narrative of shaping the earth to a new narrative of repurposing themselves in coexistence with the earth. Human history is one of progression, destruction, and reconstruction and for humanity to rethink itself back into natural world is to unearth the nature held within the body—through guises, cloaks, skins, language, foraging, and nomadic life if only individually, this is the capacity for 'wearing our ecology'. The following final chapter 'Future Human—*ultra-terrestrial worlds*' explores what a future world might look like. It explores the dystopian worlds of science fiction in which humanity did not heed the warnings and as a result plunged the earth into societal and environmental collapse whether by the fallout of nuclear war, pandemics, or off-world cataclysmic collisions. The final chapter is more a short story of the future followed by an epilogue that offers some resolution to the ideas explored in this book. One thing the final chapter does not offer is a conclusion, for to even attempt to write one would be foolish given how much is unknown and beyond the reach of any single person.

Notes

1 For a breakdown of country-by-county CO_2 emissions in 2010 and 2021, see www.statista.com/statistics/270499/co2-emissions-in-selected-countries/

2 To gather further information on the African Acceleration Adaptation Program and their climate outreach mission, see the Global Center on Adaptation website: https://gca.org/programs/aaap/

3 The Möbius strip was the invention of the 19th-century German mathematician, August Möbius. Deceptively simple, a single twist of a strip of paper gives the appearance of a two-sided object but what is in reality a continuous one-sided surface. See Thomas Nail. *The Theory of the Earth*. Stanford: Stanford University Press, 2021, p. 68.

4 Further to Bourke's Möbius strip analogy, placing human and animal on the same dialectical pathway turns to the historical argument of simulation under the disruptive evolutionary Theory of Descent developed by Darwin.

Darwinian arguments may have contributed to the deconstruction of the radical differences imagined between humans and animals, but humanism survived this attack. It did this, in part, by rejecting absolutist narratives of the human (the claim that people are utterly distinct from animals) and embracing relativistic ones (the idea of a continuum between the two states, with the fully human at one end and the fully animal at the other).

(Joanna Bourke. *What it Means to Be Human*.
Berkeley: Counterpoint, 2011, pp. 11 and 380)

5 Timothy Morton. *Being Ecological*. Cambridge, MA: MIT Press, 2018, pp. 18–19 and 107.

6 John Allen. *Topologies of Power: Beyond Territory and Networks*. Oxon: Routledge, 2016, p. 135.

7 For more information concerning the 2,000 women taking the Swiss government to the European Court of Human Rights on the basis that climate change directly impacts their lives, see www.climatechangenews.com/2023/03/29/european-court-hears-landmark-lawsuits-that-could-shape-climate-policy/

8 See William Connolly. *Facing the Planetary: Entangled Humanism and the Politics of Swarming*. Durham, NC: Duke University Press, 2017, p. 125.

9 Ibid., p. 186.

10 Ibid., p. 149.

11 For further associations concerning the human adoption of swarming in regard to military and war application, see John Arquilla and David Ronfeldt. *Swarming and the Future of Conflict.* Santa Monica, CA: National Defense Research Institute, Rand Corporation, 2000.

12 See Robert Pogue Harrison. *The Dominion of the Dead.* Chicago: University of Chicago Press, 2003, pp. 3–4. Concerning ancestry, burial, and city, see Benedict Anderson. *Buried City: Unearthing Teufelsberg, Berlin and its Geography of Forgetting.* Oxon: Routledge, 2017.

13 See Robert Pogue Harrison. *Forests: In the Shadow of Civilization.* Chicago: University of Chicago Press, 1993, p. 7.

14 Frédéric Neyrat. *The Unconstructable Earth: An Ecology of Separation,* translated by Drew S. Burk. New York: Fordham University Press, 2019, p. 45.

15 Sebald's concept of self-anaesthesia seeped into post-war German culture and literature, and it was not until the 1970s that a German perspective to experiencing destruction could emerge. The psychological absences expressed in Sebald's theory of self-anaesthesia reflected on German society consumed by guilt, whilst faced with the immense task of reconstructing its devastated cities. The enormous task of reconstruction suppressed the civilian population's suffering from the destruction of German towns and cities during the war and no voice to their experiences could be raised from a nation responsible for creating the horrors of the war and the murder of millions in death camps. See W. G. Sebald. *On the Natural History of Destruction.* London: Notting Hill Editions, 2003, p. 11.

16 Neyrat, *The Unconstructable Earth,* pp. 44 and 45.

17 Russian President Vladimir Putin's recent order to place the country's nuclear arsenal on high alert as a threat and deterrent to Western countries to desist in supplying arms to the Ukrainian military following its invasion on 24 February 2022 is viewed as no idle threat but rather as a slow build-up to humanity taking the next step towards self-annihilation.

18 See Gabriele Schwab. *Haunting Legacies: Violent Histories and Transgenerational Trauma.* New York: Columbia University Press, 2010, p. 42.

19 See Emily Martin. 'Fluid Bodies, Managed Nature'. In *Remaking Reality: Nature at the Millennium,* edited by Bruce Braun and Noel Castree. London: Routledge, 1998, p. 64.

20 The opening advertisement sequence for the NEOM Project shows a satellite view of a darkened Earth bedecked with the glow of cities, where over the horizon the promise of a rising Sun brings light as the narrator begins with the following lines: The filming sequence and narration is the stuff of a science fiction film trailer of a bright new world rising out of the ruins of the past where humanity will once again thrive and prosper. And, so the story of the NEOM Project goes, the new pet project of Saudi Crown Prince Mohammed bin Salman. To see the video advertising the NEOM Project unveiled on 28 July 2022, see www.dezeen.com/2022/07/28/video-170-kilometre-skyscraper-the-line-saudi-arabia

21 Concerning criticism of the NEOM Project, see Adam Greenfield's comments in the architectural web newsletter *Dezeen:* www.dezeen.com/2022/11/02/neom-the-line-saudi-arabia-architects-opinion

22 See Stephen Cairns. *Drifting: Architecture and Migrancy.* London: Routledge, 2004, p. 25.

23 See Simon Springer. 'Anarchism and Geography: A Brief Genealogy of Anarchist Geographies'. *Geography Compass* 7, no. 1 (2013): 56.

24 Neyrat, *The Unconstructable Earth*, p. 45.
25 Nail, *Theory of the Earth*, p. 2.
26 Ibid., p. 4.
27 Ibid., p. 15.
28 Idema points to the biological struggle in the initiation of human transmutation as a 'biological orthodoxy' with

> the idea of the environment as something external and secondary to life itself. By defining DNA as the kernel of life, genetics and genomics have consigned to the margins other subfields of biology, such as embryology and developmental biology, fields in which the environment plays a much larger role. As manifested most clearly in Richard Dawkins's book *The Selfish Gene* (1976), there is a strong ideological connection between genocentrism and anthropocentrism: isolating and controlling the genome as a kind of ultimate source of life is conducive to the notion that life on this planet now revolves around humans and their (bio)technological interventions.
>
> (Tom Idema. Stages of Transmutation: Science Fiction, Biology, and Environmental Posthumanism. New York: Routledge, 2019, pp. 7, 18, and 83)

29 Nail, *Theory of the Earth*, p. 103.
30 Jeffrey Jerome Cohen. *Stone: An Ecology of the Inhuman*. Minneapolis: University of Minnesota Press, 2015, p. 29.
31 See Helmut Tributsch. *How Life Learned to Live: Adaptation in Nature*. Cambridge, MA: MIT Press, 1982, p. 7.
32 For a chart comparison of global wealth inequality, see www.visualcapitalist.com/distribution-of-global-wealth-chart/
33 Magdoff and Williams also detail the problems of the people's revolution in the modern era, citing:

> A sustained revolutionary process, one that attains the social–ecological goals… will require the majority of the population to actively participate in the process of change. The need for a majority of the population to work toward transforming into a new society in a coordinated and organized approach raises the issue of how to overcome the staggering amount of fragmentation found among activist organizations… Older notions of solidarity are no longer as attractive to a new generation of worker–activists, trade unionists, and intellectuals. There has been an ideological assault on collective action, making it difficult to build an ongoing movement for broad social–ecological change based in factories, businesses, fields, and campuses.
>
> (Fred Magdoff and Chris Williams. *Creating an Ecological Society: Toward a Revolutionary Transformation*. New York: Monthly Review Press, 2017, pp. 405–406)

Bibliography

Allen, John. *Topologies of Power: Beyond Territory and Networks*. Oxon: Routledge, 2016.

Arquilla, John, and David, Ronfeldt. *Swarming and the Future of Conflict*. Santa Monica, CA: National Defense Research Institute, Rand Corporation, 2000.

Bourke, Joanna. *What IT Means To Be Human*. Berkeley, CA: Counterpoint, 2011.

Cairns, Stephen. *Drifting: Architecture and Migrancy*. London: Routledge, 2004.

Cohen, Jeffrey Jerome. *Stone: An Ecology of the Inhuman*. Minneapolis, MN: University of Minnesota Press, 2015.

Connolly, William. *Facing the Planetary: Entangled Humanism and the Politics of Swarming*. Durham, NC: Duke University Press, 2017.

Harrison, Robert Pogue. *Forests: In the Shadow of Civilization*. Chicago, IL: University of Chicago Press, 1993.

Harrison, Robert Pogue. *The Dominion of the Dead*. Chicago, IL: University of Chicago Press, 2003.

Idema, Tom. *Stages of Transmutation: Science Fiction, Biology, and Environmental Posthumanism*. New York: Routledge, 2019.

Magdoff, Fred, and Chris, Williams. *Creating an Ecological Society: Toward a Revolutionary Transformation*. New York: Monthly Review Press, 2017.

Martin, Emily. 'Fluid Bodies, Managed Nature'. In *Remaking Reality: Nature at the Millennium*, edited by Bruce Braun and Noel Castree. London: Routledge, 1998.

Morton, Timothy. *Being Ecological*. Cambridge, MA: MIT Press, 2018.

Nail, Thomas. *The Theory of the Earth*. Stanford: Stanford University Press, 2021.

Neyrat, Frédéric. *The Unconstructable Earth: An Ecology of Separation*. New York: Fordham University Press, 2019.

Schwab, Gabriele. *Haunting Legacies: Violent Histories and Transgenerational Trauma*. New York: Columbia University Press, 2010.

Sebald, W.G. *On the Natural History of Destruction*. London: Notting Hill Editions, 2003.

Springer, Simon. 'Anarchism and Geography: A Brief Genealogy of Anarchist Geographies'. *Geography Compass* 7, no. 1 (2013): 46–60.

Tributsch, Helmut. *How Life Learned to Live: Adaptation in Nature*. Cambridge, MA: MIT Press, 1982.

7 Future Human
Ultra-Terrestrial Worlds

Ours is indeed an age of extremity. For we live under continual threat of two equally fearful, but seemingly opposed, destinies: unremitting banality and inconceivable terror. It is fantasy, served out in large rations by the popular arts, which allows most people to cope with these twin spectres. For one job that fantasy can do is to lift us out of the unbearably humdrum and to distract us from terrors, real or anticipated—by an escape into exotic dangerous situations which have last-minute happy endings. But another one of the things that fantasy can do is to normalize what is psychologically unbearable, thereby inuring us to it. The fantasy to be discovered in science fiction films does both jobs. These films reflect world-wide anxieties, and they serve to allay them. They inculcate a strange apathy concerning the processes of radiation, contamination, and destruction that I for one find haunting and depressing. The naïve level of the films neatly tempers the sense of otherness, of alien-ness, with the grossly familiar. In particular, the dialogue of most science fiction films, which is generally of a monumental but often touching banality, makes them wonderfully, unintentionally funny.[1]

(Susan Sontag)

Imagining disasters has been just as much a part of the human psyche as imagining a world free of greed and conflict. Susan Sontag's description points to humanity's inner discord where imagination is measured in degrees of extreme calamity. Devastation is intertwined with humanity's ability to overcome its impending demise and both induce ecstasy within the viewer. We go along with the future world about to end as if it may be within our experience, even though it is not. French writer, Maurice Blanchot, who explored the philosophy of death in his work, viewed science fiction as 'an impure genre, part fiction, part science, at times extremely well-informed, at times excessively puerile'.[2] To imagine the future is to throw humanity on a course of extreme hedonism, a self-prophesizing clash of technology and psychological disorder. 'We live in a world ruled by fictions of every kind', J.G. Ballard describes in his 1973 novel, *Crash*. This is a world of

DOI: 10.4324/9781003382515-8

'mass-merchandising, advertising, politics conducted as a branch of advertising, the instant translation of science and technology into popular imagery, the increasing blurring and intermingling of identities within the realm of consumer goods, the pre-empting of any free or original imaginative response to experience by the television screen'.[3]

In Martin Rees' *Our Final Hour: A Scientist's Warning*, the author cites the 'five great extinctions' from Earth's 'geological record' that share a basis with science fiction.

> The largest of all happened at the Permian/Triassic transition around 250 million years ago; the second largest, 65 million years ago, wiped out the dinosaurs. But human beings are perpetrating a 'sixth extinction' on the same scale as earlier episodes. Species are now dying out at one hundred or even one thousand times the normal rate,

Rees warns. Past extinctions, future extinctions are both being realized in the present extinction of species today. 'Before *Homo sapiens* came on the scene, about one species in a million became extinct each year; the rate is now closer to one species in a thousand', he tells us.[4] Some species are being killed off directly, but most extinctions are an unintended outcome of human-induced changes in habitat or of the introduction of non-indigenous species into an ecosystem. In *Dystopia: A National History*, Gregory Claeys writes:

> The natural history of dystopia commences with a metaphorical Apocalypse. An angry god punishes humanity's wickedness with a flood, leaving only Noah, his family, and the animals to commence anew. It ends, seemingly, with various real dystopias created by humanity's aggression against itself and against nature.

Dystopia, cataclysmic destruction, aliens, and spaceships reaching the far extents of the universe—we have seen and talked about it without having ever been there. In the final chapter of his book, Claeys writes a fantastical portrait concerning the future of humankind and the earth where both are locked in a battle for survival. The flooding of the earth by God was 'used to regulate behaviour, to keep us morally in line, and to ensure our faith. The second represents an imminent real catastrophe. Many, perhaps most, or even all the people will perish. No Noah will save them, and no God will save Noah'.[5] Claeys posits a moral dilemma that humankind is facing in relation to biblical storytelling, the fantasy of science fiction and the reality of the earth today and trust we place in religious beliefs. 'Dystopia thus describes negative pasts and places we reject as deeply inhuman and oppressive, and projects negative futures we do not want but may get anyway', he writes. 'In so doing it raises perennial problems of human identity. Shall we be monsters, humans, or machines? Shall we be enslaved or free? Can we be 'free' or only conditioned in varying degrees? Shall we preserve our individuality or be swallowed by the collective?'[6]

How to grasp a topic so vast and threatening the existence of life on Earth humanity faces due to the effects of global warming and climate change is to look into the abyss of non-comprehension. John Berger referred to the abyss of non-comprehension between humans and animals as 'similar, but not identical'. Timothy Morton suggests that humanity's struggle is with itself, the 'hyperobject' it created—an object so vast in time and scale that its impact and effect is unthinkable or ungraspable by a human. In *Dark Ecology, For a Logic of Future Coexistence*, Morton tells '[w]hen massive entities such as the human species and global warming become thinkable, they grow near. They are so massively distributed we can't directly grasp them empirically. We vaguely sense them out of the corner of our eye while seeing the data in the centre of our vision'. Morton's 'hyperobject' is neither fiction nor science but universal. 'These 'hyperobjects' remind us that *the local is in fact the uncanny*. Space evaporates. The nice clean box has melted. We are living on a Gaussian sphere where parallel lines do indeed meet. The empty void of space and the rush of infinity have been unmasked as parochial paradigms'.[7] The 'hyperobject' is the relationship between the human species and planet Earth—something that never gets resolved for they are in constant conflict. '[W]hat hyperobjects reveal to us humans is that the whole is always weirdly *less* than the sum of its parts', he writes. The 'hyperobject' is the sum total of the fractural parts of human existence on the earth that remain out of reach of human comprehension to be joined to form a single narrative. It is not necessary to gasp the meaning of something which can be avoided—when their relatedness comes in disparate spasmolytic sequences beyond our comprehension.

Thinking about the above it could make as much as sense as not to finish this book with a fictional narrative—a science fiction imagining to what has become of human life on Earth in the year 2080. Less than 60 years from now, the year 2080 is far away enough to imagine and yet close enough in the imagination to draw on the dystopian genre of science fiction and from the perspective of human life on the earth now. Conceived as a pitch to film studio executives, the following outline for the film *Future Human* details the world in environmental and societal collapse due to ongoing human impact on the earth. The world's population is divided into two main opposing groups: the 'Technophiliacs', who believe human innovation will restore the earth while continuing to consume its resources and manufacture products; and the 'Neoecologists', who believe in the merging of human, animal, and plant life as the only way of living on the earth. A great epic struggle has emerged between the two that will decide the fate of humanity. Like science fiction films before it, the film pitch is intended to be recognizable rather than original, graspable, yet imaginable. While fictioning a future world is one way to finish this book, it avoids offering a conclusion. As such an epilogue follows to pull together the various strands explored throughout the book.

The epilogue is not intended to offer a set of conclusive answers to climate change, sustainability, and the future of human and Earth coexistence, for this is an ungraspable task for a single human to undertake. Instead, it takes the form of an afterword, looking back over what has been written to see how humans might conceive of themselves differently and the natural world around them.

Future Human—The Film Pitch

The year is 2080 and after decades of floods, drought, desertification, and acid-ification, the world's ecological system has collapsed. Inhospitable landscapes cover most of the earth and the remaining remnants of the natural world can only be found in secluded isolated pockets. As a result, humans are struggling mostly between themselves, fighting for pieces of remaining fertile land to cultivate. A vast swathe of the earth above and below the equator is unin-habitable desert where nothing lives. In the once populated regions of across Bolivia, Chile, Argentina, Brazil, and across the polluted Atlantic Ocean to vast expanses of the Sahel and Sahara deserts now encroaching Southern Europe; across to India, China, and Russia to once inhabited islands of the Pacific Ocean that now lie underwater; and onto Australia where most of the country is now desolate, Earth is incapable of supporting its population.

The failure of countries to reach agreements 50 years ago has led to the collapse of international cooperation and since early 2050 the earth has been ravished by the 30-year war between the so-called Technophiliacs and Neoecologists and it shows no signs of abating. The dominant league of humanity in terms of privilege and wealth, the Technophiliacs have placed their belief in technological solutions to solve the planet's environmental collapse while maintaining mass resources extraction, manufacturing, con-sumption, and luxury. The Neoecologists, who seek 'oneness' with the earth, on the other hand, fear such belief in technology as the means to revert cata-strophic environmental destruction, see this as merely supporting the power of industrialists, politicians, and corrupt officials. The ideologically opposed Technophiliacs and Neoecologists fight to maintain their positions and way of life on the earth, and an epic struggle for the earth's future is being fought in a never-ending cycle of confrontation while the planet sinks further down the maelstrom of environmental destruction. Money has lost all currency, whereas only gold and precious gems, which people fight and died for in illegal mines, have value. The Technophiliacs who still live with the privilege of property and luxury are traumatized to the point of being psychologically unstable from defending their domains from those who seek to take it from them. In their eyes, everyone is a potential perpetrator and people killing who threaten them is an acceptable part of living.

At the beginning of the 30-year war in 2050, tactical nuclear bombs were used by all the nations that possessed them. When one nuclear device was dropped by a rogue country, others dropped theirs in reprisal. It was always going to happen; it just needed the first despotic leader to push the button. With a smaller yield than ballistic nuclear weapons, the tactical nuclear bombs destroyed cities rather than whole countries, leaving many former capitals of the world in ruins. In the aftermath, the countries who participated in this orgy of destruction and mass killing, signed agreements pledging to 'never do it again'. All agreed that the real contest was the ideological war between them—the war of truth and lies, freedom and security, and wealth and may the best, most determined, loudest, and cruellest ideology win. Everyone accepted this course of action: it would, as some leaders decreed, 'sort out the chickens from the wolves'. With destruction behind them, the powerful governments and corporations advocated for self-protection and the abundance of guns has allowed everyone to be armed to the teeth. Regional disputes between settlers and travellers to the global conflicts between Technophiliacs and Neoecologists mean the world is engaged in endless battles.

Six decades before, when the United Nations still existed, it released a 10-point plan for humans to adopt in their fight against climate change. Illustrated in the film as a historical showreel, it shows the sequence of events for those who can remember that the 10-point plan had similarities to the information on 'how to avoid nuclear fallout', following a nuclear strike of the 1960s and 1970s. Reprised by international governments, the propaganda films from the previous century are rejigged to 'earth life' and rolled-out through government- and corporate-sanctioned media outlets. Depicting life-saving demonstrations, the films illustrate a future world of abundance in food and luxury if everyone plays their part. Children are instructed to learn programming to 'play their future' and once familiar phrases such as 'net zero carbon footprint', 'climate change', and 'green living' are outlawed in favour of the catchphrases 'tech supreme' and 'proto data'. Controlling and analysing climate information has put unlimited power in the hands of digital corporations aligned to the political elite. The decades-long ideological warfare between democracy, autocracy, theocracy, and dictatorship has sucked the life out of individual freedom and expression and enforced degrees of compliance or oppression. Religion also plays its part in redefining the world, and various campaigns undertaken by all religions seek to claim the souls of new believers known as 'the essence'. Fundamental Christian Evangelical, Islam, Hindu, Buddhist, and Taoist religions fight over the non-believers, perform ceremonial rights, and 'crusade' to attract the 'doubters' to increase their number of devotees.

Decades of global warming have led to the thawing of ice caps and bacteria strains that laid dormant for millions of years in the permafrost. These have wrought new diseases and pandemics known as 'the pest' ravages the earth's population. Rising sea levels have led to coastal cities being submerged under water, and the massive band of desert that rings the earth is referred to as 'the deadlands'. Billions of people are on the move, living in subsistence cultures and societies, and roaming bands of highway marauders known as 'the desolate' scour 'the deadlands'. The remaining tropical forests such as the Amazon and Congo Basin have been transformed into giant farmlands and open mines, and the remaining first nations peoples hold-out on the last native lands against corporate resource raiders and their private mercenary armies. Against the concerted efforts by autocratic governments and transnational corporations to spin their propaganda and make out like everything is under control, a dedicated community of hackers beam the cataclysmic truth onto billions of people's digital devices, and alliances between opposing groups are made and broken depending on the terms of survival offered. Settlements have sprung up along the superhighways that straddle the militarized boundaries of megacities selling everything from survival kits to sex. Cargo ships operate as floating cities drifting tirelessly or as ocean transporters smuggling tens of thousands of people from densely populated places to the least populated regions depending on the amount they pay.

On the human scale, the film centres on the lives of two opposing familial types of dynastic power representing the Technophiliacs and Neoecologists and the steps they take to keep their authority. What might appear as a good versus evil scenario: where consumption, privilege, and power of technology are pitted against universal freedoms and environmental preservation is in fact the failure of humanity to move beyond duopolies. Technophiliacs and Neoecologists have taken to protecting their cities and communities through violent military enforcement in-keeping with old-age agenda of nationalist and racial segregationist values that has plagued humanity for centuries. Technophiliacs have relieved humans from performing manual labour via automation; 'freedom time' is enshrined as 'the way' of modern life. Artificial intelligence and surveillance control most aspects of society including law enforcement and human behaviour. Decades previous, the Neoecologists revolted from this repressive authoritarianism and in their billions took to the countryside to form loosely knit societies, compiled seed banks, and built greenhouses that filtered clean air in which to grow their food. They talk of inclusiveness but given the extent of violence, exclusion is the name of the game they defend their fortified settlements from roaming bands of 'the desolate' at the hand of the gun. Imitating ancient citadels and fiefdoms, these large groups control their domains and grant citizenship known as 'the allegiance'. Ongoing battles fought over decades between the rival ideologies, rogue states, and clans dominate life on the earth. Multi-ethnic groups of young people who, having rebelled against their parents and the authorities who they see as having failed to take action when they had the chance, grow in numbers forming tribes collectively referred to as 'the terrorists'. Societies across the globe now reflect the earth's climatic condition; they are societies of disruption and self-destruction, fragility, and decline.

Over the last decades, once abundant animals roaming across the world have declined with many species now extinct and to capitalize on people's insatiable appetite for entertainment, animal circuses better known as 'the Ark' travel the world. For in-home entertainment, animals such as elephants and giraffes and apex predators such as lions and leopards that once roamed the African continent are mythologized as avatars for the furtive imaginations of children in bedtime stories. The depletion of native forests and plant species once numbering in the millions now number in the thousands, which has led to a serious decline in plant photosynthesis to absorb CO_2 from the atmosphere while also devastating insect populations, especially the bees needed to cross-pollinate. Food varieties are reduced to basic items that are propagated from genetically modified seeds grown in huge greenhouse laboratories stretching kilometres in length, and vast factories manufacture vitamin supplements necessary for human health to those who can afford it. Famine outbreaks are common, causing conflict, starvation, and death.

Many cities not destroyed during the nuclear conflicts 30 years ago have reached gigantic proportions with each divided along generic lines of demarcation separating the monoculture minority of the privileged from the multitude ethnicities of the majority. Divisions between developed and underdeveloped peoples across the world remain as they had always done, and the scourges of inequality and poverty have increased. Yet, some gains have been made in pockets of countries where people have risen to take back their lands by expelling transnational corporations and preventing them from plundering resources and stopping corrupt politicians from profiteering though no-one escapes their reach.

Long periods of grey sky, thick with particles, cloak the atmosphere in a choking mess of CO_2-covering cities and regions before being blown onto new regions. Known as 'the dark', once blown away by cyclonic winds, the 'big blue' temporarily reappears and celebrations are held. The long-standing divisions between the 'Global North' and 'Global South' have been erased by the unending catastrophic weather events and now a great human migration of the 'world weary' travel the earth. No-one is surprised by how far humanity has fallen, in fact it has become acceptable, palatable, desirable even. For some, this dystopian world brought about new opportunities, a new wild west frontier at the end of a gun, while others are fed the ultimate escape with the possibility of leaving Earth to off-Earth planetary worlds. There is no time to be depressed or yearn for a better world. The climax at the end of the film is an epic battle between the Technophiliacs and Neoecologists and many die. The dynasty of the two powerful family types that ruled have been erased and, in their place, the possibility of a new world is reprised and the possibility for a sequel. With the pitch over, the film executives leave with their final comment asking for more explicit scenes of violence, destruction, and sex.

There is a reason for this unoriginal film pitch: as mentioned, I thought I could offer no conclusion to this book and fictionalizing science for the future of the world seemed to offer a way out. As science fiction depends on recognizing human behaviour and language for its understanding and communication to project a fictive future scenario, it made sense to me to run the themes in this book to their endpoint, that is write their science fiction. When

I submitted my original book treatment, a reviewer suggested that if I were to write a summary in this 'unscientific' path that breaks somewhat with the formal expectations of academia, it would jeopardize any real effectual meaning this book may offer to those wanting to know more about climate change and sustainability. In response, further I have written an epilogue to collect my ideas. If I were to write a conclusion even though I know it is not possible to offer one, it would be to say that eventually everyone, wherever they live, will become a refugee of climate change.

Epilogue—Factual Reality

The various themes raised throughout this book have sought to compile a critique, develop approaches, and rethink what we know and practice concerning sustainability in response to the human impact of climate change. The book has raised issues concerning human impact on the earth, eliminating fossil fuels, reducing resources extraction, overhauling the domination of capital, removing human inequality, confronting post-colonial histories, dismantling corporate imperialism, supporting climate vulnerability and displacement, abiding by weather, animal and plant species for an emerging environmental connectivity, to international agreements on sustainability, decarbonization technologies, and climate activism in addressing what becomes of human life on Earth in the 21st century.

The concept of 'wearing our ecology' formed a backbone to the book as a way for humanity to become inseparable from the natural environment as the foundation of living with climate change. It proposed that human separation from the natural environment came as a consequence of humans dominating the natural world, viewing its resources solely in terms of how they might benefit humankind. The book described how this separation began in the transition from nomadic life to the establishment of settlement, the industrial revolution, and the technologies of mass production and consumption. The book also proposed ideas as to how humanity might rethink itself and redefine its 'nature' in reference to other life forms such as animal and plant society, transmutation, and metamorphosis to formulate new ways for environmental adaptation in light of the effects of climate change. It also noted the roles that religion has played in how humans view their relationship to the natural world. The differences between Christian, Islamic, Hindu, Buddhist, and multiple indigenous beliefs concerning the natural world are fundamental to understanding human impact on the earth. Early European colonial powers used Christianity to forcibly dispossess first nations peoples from their lands from the 16th to the 20th centuries, and this has directly led to them being vulnerable to climate change in the 21st century. The role of cartography during colonialism was crucial, where lines of demarcation dissected native peoples' cultural and geographical connections to land into artificial borders of made-up countries. The book considered what role cartography might play in the 21st century in mapping global routes for climate refugees in response to climate change. It also looked at the role of borders and nationalism in terms of the territorial protectionism of nation-states in generating fear and threat in Western countries to deny freedom of movement of climate change refugees fleeing their homelands due to global warming not of their making. It explored how extreme weather events have become mediatized entertainment as the world lurches from one catastrophic weather event to the next. One of the recurring arguments made was that there can be no global sustainability without global equality. What seems obvious is in fact fast becoming more remote as global inequality continues to increase.

The book has sought to combine history with the present and the future to illustrate how humanity got to the point of seeding its own destruction and how it is now desperately trying to save the earth from catastrophic environmental collapse in order to save itself. In large part, the book has focused on the human costs of climate change and how investments in untested technologies are giving false hope in solving the world's environmental problems rather than pursuing the path of a complete reordering of human social structures, capital, and habitation on the earth. The book also took into account how the impacts of climate change are gendered, affecting women and girls differently especially those living in vulnerable regions responsible for agricultural food production and water collection. The book is not intended to overwhelm the reader by pointing to the negative position in which humankind finds itself, but rather to 'unearth' new positive attitudes to harness the future preservation of the earth's natural habitats and ultimately humankind.

To write the epilogue of this book is to start at the beginning in the introduction which outlined the impossible task of tackling the most pressing issue facing humankind—that is, its survival in a climatically unstable, turbulent world. The introduction made no claim that this book will affect how humans live on Earth, how governments and institutions act with greater responsibility, or how fossil fuel companies and global mining corporations reduce their operations and impact on the climate. Humanity's relations with the natural environment from nomadic life to the establishment of settlement to the industrial age and on to the globalization today have been bent into unrecognizable form. The idea of presenting a future human society based on terrestrial mobility and climate adaptation—where we can 'wear our ecology'—means excavating the buried 'nature' of human nature in securing a viable world for future generations. The introduction cited the overwhelming amount of information generated daily concerning climate change events and global sustainability policies, which was impossible to keep up with—what was current one day was obsolete the next. The introduction outlined the themes of each chapter and upon looking at the titles it would seem they point to a progression of my thoughts and future directions for the area of sustainability.

Chapter 1, 'Sustainability's Paradox', focused on the effects of climate change on billions of people throughout the world, where at any one time 300 million people seek refuge, 2 billion people experience food and water insecurity, while the richest 1% of the world's population produce more carbon emissions than the whole of Africa combined. It pointed out that if humanity continues to fail to secure the basic human rights of food security, health, education, gender equality, work, and freedom of mobility to all people, sustainability will fail. In a world where commodities appear to have more rights and access to the world than many humans, sustainability will not be reached. The chapter cited various UN climate change COP conferences, NGO programmes, climate activist campaigns, and commercial projects from released reports and media coverage.

Chapter 2, 'Terrestrial Migrations', looked at human migration in relation to the physical embodiment of nomadic people through to the virtual access of global telecommunications. At one end digital technologies open the world to radically change our knowledge of the world, revealing the immense differences and desires between people. The chapter explored how real-time representations of the world, if only 'pictorially', also reproduce the cruelty of a value system where visual access to the reality of migration restrictions disfigures/distorts the representation of the world for the most vulnerable. It highlighted the extreme immorality of repelling climate change migrants based on the colour of their skin, religious beliefs, customs, and culture, or seeing them as a threat to Western societies—all of which does nothing to help realize a sustainable world. To make climate change racial in this way is abhorrent and perpetuates the racial inequality that has been institutionalized over centuries.

Chapter 3, 'Earth Extractions', explored human relations with the natural environment as being based solely on how nature benefits humanity. Digging our way through the earth's resources, felling forests, rerouting water sources to service energy needs, creating wealth by further degrading the environment—this is what has characterized human progress: a blind development at the expense of the earth. The systemic dismantling of ecologies, regions, and their people's ability to survive devastating climate events due to land degradation is a clear result of uncontrollable human impact. Where populations around the world are suffering food and water scarcity, drought and famine due to climate change, rich Western countries by contrast suffer the least, a situation that is, in the words of UN Secretary General Antonio Guterres, both 'immoral' and 'untenable'.

Chapter 4, 'Weathering Patterns', considered how weather has suffered the fate of humanity's indifference through its ability to dissect and disassociate its elements through building—dividing experience between the material surfaces of inside and outside. The chapter explored relations between weather, animals, plants, and humans to suggest that the separations that have been built up over time are no longer viable in an era of extreme climate change. The chapter formulated the analogy where humanity becomes animal and plant, bringing together these three societies at a deeper level, so as to shift humans away from the central place that they have occupied on the earth in favour of an interconnected, inseparable co-occupation. For human and Earth coexistence to endure, weather has to be worn, as happened in early nomadic life, and city and rural populations must be connected so that the effects of climate change are better understood and shared. The chapter suggested that through a collective osmosis, humans can rework themselves into animal and plant life as equal 'citizens' of the earth. To start this journey, humanity must reject its 'nature' of rampant consumption, assume new economic and social models, and stabilize the earth's weather patterns so a future sustainable world can emerge.

Chapter 5, 'Climate Gathering', sought to combine human and Earth coexistence through the notion of 'wearing our ecology' based on humans' co-dependency with the natural world. In the era of climate emergency, the chapter proposed that the cartographical spatial delineations of nation-states be dissected, boundaries dissolved, artificial surfaces of the city made porous, and connections between rural and urban populations bridged for sustainability to be a global reality. It proposed that a new cartography of climate change will begin to emerge across the continents, foregrounding a new geo-spatial global order. The mass of criss-crossing lines illustrating the mobility of climate refugees will stabilize to forge a new nomadic resettlement of the earth where individuals and collectives operate independently of established authorities to devise their own ecological societies based on living with the least resistance with the earth. The chapter proposed that 'wearing our ecology' requires a far greater connectedness between humans and their surroundings than what presently exists, opening up new opportunities for urban and rural transformation. It also means the removal of global divisions such as the 'Global South' and 'Global North'—a terminology that only reinforces inequality. Adaptability is the key but in order to achieve this, humanity must learn to become its ecology, to comprehend itself as its surroundings.

Chapter 6, 'Environmental Adaptations', looked at how to construct human adaptation and variation in creating departures for a new coexistence with the natural world. The idea of human transformation in adapting to climate change is not intended to echo Darwinian theory by suggesting that humans have the same capacity for transmutation as animals or plants; instead, it is to be in accordance with nature. For sure, the chapter is speculative, lacking descriptions in how to achieve this and giving no guarantees that humanity can arrest the extreme weather turbulence it has created. Human transmutation does not involve a remodelling shaped by its surroundings but rather connectivity to its natural genetics. The proposed idea is to radically alter government and the power of capital in favour of people-led societal orders based on sustainable living, which requires a fundamental shift in human perception. Simple as it is complex, it nevertheless remains unapproachable under the current global system. Humanity is still dealing with age-old issues that have scuppered its progress, such as violence, terror and war, dictatorship, corruption, and greed and most visibly economic, racial, and gender inequality. Yet, there is evidence that human transformation is underway in people's awareness, eating habits, consumption, and waste management, showing that humanity is willing to exchange some privileges and luxuries for climate security. The chapter asked how humanity might rethink itself back into the natural world and to unearth the 'nature' held within the human body—guises, cloaks, skins, language, foraging, nomadic life, etc., all formulate the environmental adaptation needed to realize the goal of 'wearing our ecology'.

This book has charted the moves and shifts, the forward and backward steps humans have taken so far in attempting to tackle climate change. The human dimension of this undertaking remains inconsistent—inequality between rich and poor, histories of colonialism, racial discrimination, resource expropriation, religious intolerance, gender disparity. Addressing these inconsistencies is fundamental to achieving global sustainability. The book has referenced a broad spectrum of works on global warming, climate change, sustainability, adaptation, ecology, the biosphere, environmental science, meteorology, urbanism, migration, human geography, colonialism, capitalism, and corporate imperialism, as well as works on botany, plants, and animals. But of course, there are many more areas to address and I feel I have reached my capacity. To repeat, in undertaking this book, my aim was not to offer a definitive guide of how and in which direction humanity should proceed in trying to find its way out of the environmental devastation it has wrought. I consider that in order to achieve global sustainability, it is necessary to understand humanity's past domination of the natural world as a prerequisite for reforming its present psychological dislocation and reshaping its interaction with nature in the future. For sure, this is the only way in which future generations can be guaranteed that they will inherit a stable, livable Earth.

Notes

1 Susan Sontag. *Against Interpretation*. New York: Picador, 1965, pp. 224–225.
2 Maurice Blanchot. 'The Proper Use of Science Fiction'. In *Imagining the Future: Utopia and Dystopia*, edited by Andrew Milner, Matthew Ryan, and Robert Savage. North Carlton: Arena Publications Association, 2006, p. 376.
3 See J.G. Ballard. *Crash*. New York: Picador, 2001, p. 48.
4 Martin Rees. *Our Final Hour: A Scientist's Warning*. New York: Basic Books, 2003, pp. 102–103.
5 Gregory Claeys. *Dystopia: A National History*. Oxford: Oxford University Press, 2017, p. 497.
6 Ibid., p. 489.
7 Timothy Morton. *Dark Ecology, For a Logic of Future Coexistence*. New York: Columbia University Press, 2016, p. 11.

Bibliography

Ballard, J.G. *Crash*. New York: Picador, 2001.

Blanchot, Maurice. 'The Proper Use of Science Fiction'. In *Imagining the Future: Utopia and Dystopia*, edited by Andrew Milner, Matthew Ryan, and Robert Savage. North Carlton: Arena Publications Association, 2006, pp. 375–383.

Claeys, Gregory. *Dystopia: A National History*. Oxford: Oxford University Press, 2017.

Morton, Timothy. *Dark Ecology, For a Logic of Future Coexistence*. New York: Columbia University Press, 2016.

Rees, Martin. *Our Final Hour: A Scientist's Warning*. New York: Basic Books, 2003.

Sontag, Susan. *Against Interpretation*. New York: Picador, 1965.

Index

For Product Safety Concerns and Information please contact our EU representative GPSR@taylorandfrancis.com
Taylor & Francis Verlag GmbH, Kaufingerstraße 24, 80331 München, Germany